DAVID BOSWELL REID

TEXT-BOOK

FOR

STUDENTS OF CHEMISTRY

CONTAINING

A CONDENSED VIEW

OF THE

FACTS AND PRINCIPLES OF THE SCIENCE

libron Classics
www.elibron.com

Elibron Classics series.

ISBN 1-4021-6415-7 (paperback)
ISBN 1-4212-8091-4 (hardcover)

This Elibron Classics Replica Edition is an unabridged facsimile of the edition published in 1836 by Maclachlan and Stewart, Edinburgh.

Third Edition, 8vo, with Eighteen Engravings, price 21s.

5. **A COMPENDIUM of the DISEASES of the HUMAN EYE,** to which is prefixed an account of the Anatomy and Physiology of that Organ. Illustrated with Eighteen Plates, partly coloured. By A. WATSON, F.R.C.S.E., &c.

"Mr Watson's work exhibits a systematic and comprehensive view of the diseases of the human eye. The descriptions are short but very distinct; and as they are manifestly derived chiefly, if not entirely, from personal experience, they shew a talent for observation and pathological acuteness by no means common."—*Edin. Med. and Surg. Journ.* No. 96.

Second Edition, in one vol. 8vo, price 15s. boards,

6. **ELEMENTS of PRACTICAL CHEMISTRY,** comprising a series of Experiments in every department of Chemistry, with directions for performing them, and the preparation and application of the most important Tests and Re-agents. By D. B. REID, M.D., Fellow of the Royal College of Physicians, and Lecturer on Chemistry, &c.

Second Edition, in one vol. 8vo, price 6s. sewed,

7. **TEXT-BOOK for STUDENTS of CHEMISTRY,** comprising a Condensed View of the Facts and Principles of the Science. By D. B. REID, M.D., Lecturer on Chemistry, Fellow of the Royal College of Physicians, formerly Senior President of the Royal Medical Society, &c. &c.

Second Edition, with Letterpress, price 6s.

8. **SCALE of CHEMICAL EQUIVALENTS,** with Directions for using it; and a short Explanation of Definite Proportions. By D. B. REID, M. D.

In 12mo, with Plates and Woodcuts, price 9s.

9. **OUTLINES of MEDICAL BOTANY,** comprising Vegetable Anatomy and Physiology, the Characters and Properties of the Natural Orders of Plants, an Explanation of the Linnean System of Classification, and Tables of Medicinal Plants, arranged in their Linnean and Natural Orders. By HUGO REID.

"Decidedly the best introductory work on the subject."—*Medico-Chirurg. Review.*

"A very good little book, well timed, and well executed."—*Medical Gazette.*

"An elementary work on Botany, for the use of Students of Medicine, for which purpose it is well adapted."—*Edin. New Phil. Journ.*

In 12mo, price 7s. 6d.

10. **MANUAL of OPERATIVE SURGERY,** translated from the Third Edition of the French of J. COSTER, M.D. By G. FYFE, M.D., Extraordinary Member of the Royal Medical Society.

In one vol. 8vo, with Coloured Plates, price 42s.

11. **ALGÆ BRITANNICÆ,** or Descriptions of the Marine and other Inarticulated Plants of the British Islands, belonging to the order of Algæ, with Plates illustrative of the Genera. By ROBERT K. GREVILLE, LL.D., F. R. S., & F. A. S. E., &c.

In six vols. royal 8vo, the Plates beautifully Coloured, price L. 15,

12. **The CRYPTOGAMIC FLORA,** or Coloured Figures and Descriptions of Cryptogamic Plants belonging chiefly to the order of Fungi, and intended to serve as a continuation of English Botany. By R. K. GREVILLE, LL.D., F. R. S. & F. A. S. E., &c.

In royal 18mo, the Tenth Edition, just published, price 8s.

25. GREGORII CONSPECTUS MEDICINÆ THEORETICÆ. Auctore JOCOBO GREGORII, M. D. Olim. Med. Theor. Nuper Med. Pract. in Acad. Edin. Prof.

In 18mo, price 2s. boards,

26. THE FAMILY MEDICAL GUIDE. By J. MACKENZIE, M. D., Fellow of the Royal College of Physicians, &c. &c. Edinburgh.

In one vol. 12mo, price 9s. boards,

27. DESCRIPTIONS of the RAPACIOUS BIRDS of GREAT BRITAIN; comprehending the Eagles, Hawks, and Owls; with Engravings on Wood and Copper, illustrative of the Genera, and the Modifications of the External and the Intestinal Canal. By WM. MACGILLIVRAY, A. M., Conservator of the Museum of the Royal College of Surgeons in Edinburgh.

In 8vo, price 2s. 6d.

28. The PHRENOLOGICAL JOURNAL and MISCELLANY. Published quarterly.

In 12mo, price 4s., fifth edition, and also in royal 8vo, price 1s. 6d.
"The People's Edition,"

29. The CONSTITUTION of MAN, considered in Relation to External Objects. By GEORGE COMBE.

In 8vo, price 1s.

30. OUTLINES of PHRENOLOGY, Illustrated by Engravings on Wood. By GEORGE COMBE.

In 8vo, price 1s. 6d.

31. LECTURES on POPULAR EDUCATION. By GEORGE COMBE.

In 8vo, price 1s. 6d.

32. A LETTER from GEORGE COMBE to FRANCIS JEFFREY, Esq., in Answer to his Criticism on Phrenology in the 88th Number of the Edinburgh Review.

In 8vo, price 2s. 6d.

33. TESTIMONIALS in favour of PHRENOLOGY, as the SCIENCE of MIND; presented by GEORGE COMBE on becoming a Candidate for the Chair of Logic in the University of Edinburgh.

34. EDINBURGH UNIVERSITY ALMANACK. Price 2s. 6d.

In one volume 8vo, price 18s., the Sixth Edition,

35. The FIRST LINES of the PRACTICE of SURGERY. By SAMUEL COOPER, Professor of Surgery in the University of London, &c. &c.

In the Press.

A SYSTEM of PHRENOLOGY. By GEORGE COMBE. In two volumes 8vo, Fourth Edition, with numerous Cuts.

OUTLINES of MORAL PHILOSOPHY, founded on Phrenology. By GEORGE COMBE.

ELEMENTS of PHRENOLOGY. By GEORGE COMBE. Fourth Edition. 12mo.

SELECTIONS from the PHRENOLOGICAL JOURNAL; comprising Forty Articles in the first Five Volumes, chiefly by GEORGE COMBE, JAMES SIMPSON, and ANDREW COMBE, M. D. Edited by ROBERT COX. In one neat vol. 12mo.

TEXT-BOOK

FOR

STUDENTS OF CHEMISTRY.

John Johnstone Sc.
94 South Bridge

SKETCH OF THE ARRANGEMENTS IN D^r. D. B. REID'S CLASS ROOM,

ROXBURGH PLACE, EDINBURGH.

TEXT-BOOK

FOR

STUDENTS OF CHEMISTRY:

CONTAINING

A CONDENSED VIEW

OF THE

FACTS AND PRINCIPLES OF THE SCIENCE.

BY D. B. REID, M. D., F. R. S. E.,

LECTURER ON CHEMISTRY,

FELLOW OF THE ROYAL COLLEGE OF PHYSICIANS, FORMERLY SENIOR PRESIDENT OF THE ROYAL MEDICAL SOCIETY.

SECOND EDITION,

ENLARGED AND REVISED

EDINBURGH:

MACLACHLAN AND STEWART, SOUTH BRIDGE.

MDCCCXXXVI.

PRINTED BY NEILL & CO. OLD FISHMARKET.

PREFACE.

ALTHOUGH the student is already supplied with many excellent Elementary Treatises on Chemistry, there appears still a want of a summary such as the present to prepare him for lectures, to serve as a NOTE-BOOK, and to enable him afterwards easily to recall his knowledge, and frequently to recapitulate, as he advances in the study of the science. The precision of chemical language, and the use of signs and diagrams, of which I have freely availed myself, assist much both in condensation and clearness in a work of this description, consisting chiefly of brief and condensed notices of leading facts and principles.

I am aware, that, in a science every step of which is illustrated by experiment, books, however valuable in many respects, are very secondary to experimental demonstrations; the truth of the poet's observation, "Segnius irritant animos demissa per aurem, quam quæ sunt oculis subjecta fidelibus," applies to Chemistry above all other sciences. But a full exposition of facts and principles must still be valuable to the student, if properly adapted to his capacity and progress, leading him step by step from what is simple to what is complex,—taking nothing for granted which should be proved,—and putting him in possession of the materials for judging before a conclusion is drawn, so that his knowledge becomes real, and rests on a proper foundation.

The changes necessarily arising from the progress of a science which, in its extraordinary career, now illustrates the whole economy of Nature, and guides or improves almost every process of art, have produced a revolution in the system of chemical teaching, and imposed new duties on the teacher, which necessarily engross a much larger share of his time and attention. Indeed, it would be strange if the system of teaching a science of such recent origin and rapid progress were not susceptible of improvement, while its methods are becoming daily more complete, its instruments more perfect, varied, and powerful, and its practical applications more important. Besides, the juster views that are now entertained of the nature and objects of education, require the remodelling of a system which had its origin in times of comparative darkness, and when Chemistry was a very subordinate art.

Increased attention to the practice of Chemistry, and its adaptation to the progress of society, form perhaps the most important features this science presents in modern times; so that even those whose principal object is a knowledge of the Theory of Chemistry, can now attain this with most facility by familiarity with practical operations. *From the improvements in the method now adopted for experimenting on the small scale, every individual who has once learned this branch of chemical manipulation, is enabled to repeat the greater number of the experiments which he sees in the lecture-room, again and again, at an expense so extremely moderate, compared with what is generally supposed to be necessary, that it may be truly said, that it is now within the reach of every one to study Chemistry practically at home as well as at the class-room.*

In endeavouring to promote the means of attaining so desirable an object, the tube-apparatus, so ably described by Dr Faraday, and the system of testing on a small scale, for which chemical science is so much indebted to the late Dr Wollaston, may be adopted. But, for the greater number of experiments, I have been anxious for some time past to introduce the use of the fragments of window glass, which may be obtained at any glazier's, having found that with these I have been enabled to render Practical Chemistry more accessible to all classes of the community who may wish to operate at home with a simple and economical apparatus (as well as to introduce this department of science as an early branch of general education in schools and academies); *for, on these fragments of glass, when properly selected, it is easy to boil fluids, dissolve solids, filter and crystallize solutions*, and even to *conduct a great variety of furnace operations* on a small scale, with the assistance of a good lamp. The rapidity with which experiments may be made in this manner, is another great advantage which this kind of apparatus presents. It is described under the title of FLAT GLASS APPARATUS*

* "Though a complete and extended course of chemical experiments can be given only where the proper facilities for operating are to be met with, the introduction of a system of testing and operating upon a small scale, as a means of facilitating the progress of the student, is of the greatest practical importance. He who provides himself with a few slips of glass, a few phials with acids and alkalis, and a small quantity of fifty or sixty of the more important chemical preparations —all of which, both apparatus and materials, he can procure for five or six shillings at the utmost—not only provides himself with a small museum to which he

in my RUDIMENTS OF CHEMISTRY. But it is scarcely necessary to add, that experiments conducted in this manner, however useful and inte-

can continually refer, but can also perform, again and again, with a part of the materials, several thousand experiments. He can, moreover, renew them at his own convenience, whenever it may be desirable; he can study them leisurely at home with his Text-Book before him, and repeat those which may be more complicated, till he becomes familiar with them.

The small museums of tests and specimens, prepared by Mr Macfarlane of Edinburgh, are arranged in the manner I have recommended. They contain about sixty specimens of the most important chemical preparations, including a few phials with acids and alkalis, a test-tube, slips of flat glass, filtering paper, and test-paper. These alone, independent of the larger and more complete portable laboratories and chemical test-boxes, cannot fail to be of the highest value to the student of chemistry. I do not hesitate to affirm that he who provides himself with such specimens, and learns the method of operating with flat glass, will acquire, by frequently testing his materials at home, an available knowledge of the science, and a specific recollection of the leading truths it presents, which he who contents himself with what he sees in the class-room can never attain. However important the demonstrations and exercises there may be, still the student is too often apt to imagine that he is perfectly familiar with the subject, because he can follow the connected exposition given by the teacher; but when he begins alone, and is left entirely to his own resources at home, then he soon finds out what he does know and what he does not, and is put in the train of mentioning his difficulties to his teacher, and paying a more close attention to every thing that is brought before him.

In using these materials, each of the sixty specimens can be exposed on the flat glass to the action of heat, of water, of acids, of alkalis, and of each other; so that, before entering upon any complicated mixture, the student can easily try thousands of experiments. If any one contrast the progress of a student who has been accustomed to operate in this manner, with that of another who has never availed himself of this means of information, who has only seen a specimen in a museum or lecture-room, instead of having the opportunity of recurring to it, and operating with it as often as he may wish, he will at once perceive the benefit that accompanies this frequent repetition of experiments on the small scale, each of which may present phenomena as palpably to the eye as they are observed in the demonstrations of the lecture-room. Those who have not witnessed the amount of instructive and interesting experiments which can be performed by all the pupils in any school or academy, with the most simple means, where hundreds may be operating at the same moment, cannot appreciate the zeal and enthusiasm with which the students enter upon the experimental department, the practical benefit which they derive from such instruction, and the mental training which necessarily accompanies it. Nor is it a matter of small importance, that the adaptation for educational purposes of those modes of operating on a small scale, which analytical chemists have so successfully cultivated, promises to be equally useful to the professional student, to those who study science as an intellectual pursuit, and to the artisan who may require to test the various materials with which he works. Of late, indeed, I have seen the slips of glass used in a chemical manufactory by the workmen in testing the materials which they employed."—See "*The Study of Chemistry, its nature and influence on the progress of Society.*"

resting, can never supply those resources, and that knowledge of the practice of this science, which is to be obtained only by operating with the various kinds of apparatus which the progress of Chemistry has rendered necessary in an extended Practical Course, the great importance of which has now been so generally admitted by those Public Bodies who have rendered it imperative for different classes of students to attend to the Practice of Chemistry.

In the selection of Symbols, I have adopted, as far as I could, the system of Berzelius: dots representing oxygen are placed before instead of above the symbols, being much more convenient in printing. In the table of Elementary Equivalents, I have stated the more important alterations proposed lately, but have not thought it necessary to incorporate them in the text, until farther experience shall confirm them, more especially as nothing has of late proved so embarrassing to the student as a frequent change of equivalent numbers. It is to be hoped that arrangements will soon be concluded in which all the chemists in this country will join, in regulating the nomenclature and symbols to be employed, and also in regard to the equivalent numbers it is most desirable to adopt.

CONTENTS.

CHEMICAL EQUIVALENTS AND SYMBOLS.—HYDROGEN = 1.

(Fractions are omitted in this Table.)

Symb.		Equiv.	Discovered by	Year.
H.	Hydrogen	1	Cavendish,	1766
C.	Carbon	6	Known to the Ancients.	
O.	or Oxygen	8	Priestley, / Scheele,	1774 / 1775
B.	Boron[1]	8	Davy,	1807
N.	Nitrogen	14	Rutherford,	1772
S.	Sulphur	16	Known to the Ancients.	
P.	Phosphorus	16 ?	Brandt,	1669
F.	Fluorine	18 ?	Dr Faraday	1834
Cl.	Chlorine[2]	36	Scheele,	1774
Se.	Selenium	40	Berzelius,	1688
Br.	Bromine[3]	75 ?	Balard,	1827
I.	Iodine[4]	124	Courtois,	1811
Si.	Silicum	8	Berzelius,	1824
L.	Lithium	10	Arfwedson,	1818
Al.	Aluminum[5]	10	Wohler,	1828
Mg.	Magnesium	12	Bussy,	1829
G.	Glucinum	18	Wohler,	1828
Ca.	Calcium	20	Davy,	1807
Na.	Sodium[6]	24	Davy,	1807
Co.	Cobalt[7]	26	Brandt,	1773
Ni.	Nickel[8]	26	Cronstedt,	1751
Fe.	Iron	28	Known to the Ancients.	
Mn.	Manganese	28	Gahn and Scheele,	1774
Cr.	Chromium	28	Vauquelin,	1797
Ir.	Iridium	30	Descotils, / Tennant,	1803
Ti.	Titanium	32	Gregor,	1791
Te.	Tellurium	32	Muller,	1782

Symb.		Equiv.	Discovered by	Year.
Cu.	Copper	32	Known to the Ancients.	
Y.	Ittrium[9]	34	Wohler,	1828
Z.	Zinc[10]	34	Known to Agricola,	1529
As.	Arsenic	38	Brandt,	1733
Zr.	Zirconium[11]	40	Berzelius,	1824
K.	Potassium	40	Davy,	1807
Sr.	Strontium	44	Davy	1807
Sb.	Antimony	44	Basil Valentine,	1490
R.	Rhodium[12]	44	Wollaston,	1803
Mo.	Molybdenum	48	Hielm,	1782
Ce.	Cerium[13]	50	Hisinger, / Berzelius,	1804
Cd.	Cadmium	56	Stromeyer,	1818
Pd.	Palladium	56	Wollaston,	1803
Sn.	Tin	58	Known to the Ancients.	
Th.	Thorinum[14]	68	Berzelius,	1829
Ba.	Barium[15]	70	Davy,	1807
Bi.	Bismuth	72	Known to Agricola,	1530
W.	Tungsten	96	D'Elhuyart,	1781
Pt.	Platinum[16]	96	Wood,	1741
Pb.	Lead	104	Known to the Ancients.	
Ag.	Silver[17]	110	Known to the Ancients.	
Ta.	Columbium	143	Hatchett,	1802
Hg.	Mercury[18]	200	Known to the Ancients.	
Au.	Gold[19]	200	Known to the Ancients.	
U.	Uranium	208	Klaproth,	1789
Os.	Osmium	100	Tennant,	1803
V.	Vanadium	67	Sefstrom,	1830
D.	Donium	?	Richardson	1836

Alterations proposed in equivalent numbers.

[1]11. [2]35. [3]78. [4]126. [5]14. [6]23. [7]29. [8]29. [9]32. [10]32. [11]33, [12]52. [13]46. [14]60. [15]69. [16]99. [17]108. [18]202. [19]196.

EARTHS discovered before Davy's discovery of the Composition of the Fixed Alkalis and Earths, and not known to the Ancients.

Baryta	by	Scheele,	1774
Strontia		Hope,	1792
Zirconia,		Klaproth,	1789
Glucina,	by	Vaquelin,	1797
Thorina,		Berzelius,	1828

1. Each dot (.) prefixed to a symbol indicates one equivalent of oxygen.

2. Compounds are represented by the symbols of their elements. SFe = sulphuret of iron.

3. A figure prefixed to any symbol or symbols, multiplies all that follow, unless a new sign intervene. 2·H⁚S = 2·H+2⁝S. 2·H⁝S+·K = 2·H+2⁝S+·K.

4. A small figure placed after any symbol indicates that it is to be multiplied,—and it does not refer to any other symbol. Thus: $H^2C = 2H+C$. $C^2N = 2C+N$. $NH^3 = N+3H$. $OC^2H^3 = O+2C+3H$.

5. Brackets are often used in representing complicated combinations. $(2HC^2N+C^2NFe)$+2·K = Ferroprussic acid + 2 Potassa.

6. A figure prefixed to any symbols placed between brackets, multiplies them all, whether signs intervene or not.

7. In the Symbolic and Diagrammatic Illustrations, " & " is used to signify a substance added to or separated from another, but + is placed between substances chemically combined. Thus ⁝S&·K signifies that sulphuric acid is to be added to the potassa, but ⁝S+·K means sulphate of potassa, or sulphuric acid already combined with potassa.

8. Many vegetable acids and other substances are represented most conveniently by *Italic* Capitals, adding some succeeding letter where more than one have the same initial letter.

A. Acetic.	*Cz*. Carbazotic	*F*. Formic	*M*. Malic	*S*. Succinic
B. Benzoic.	*Ct*. Citric	*G*. Gallic	*Mc*. Meconic	*T*. Tartaric
C. Camphoric.	*Cy*. Cyanic	*K*. Kinic	*R*. Racemic	

EXERCISES ON SYMBOLS.

To the left of the sign (=) are placed the materials used; and to the right, the products from their reaction.

·H&Fe = ·Fe&H
·H&Fe&⁝S = ⁝S·Fe&H
·H&Z&⁝S = ⁝S·Z&H

:Hg = Hg&O^2
2:Mn = ⁝Mn^2&O
:Mn&⁝S = ⁝S·Mn&O
:⁝Cl·K = ClK&O^6

5·N^2&P^2 = :⁝P^2&10N
4·N^2&SFe = ⁝S·Fe&8N
NH^3&3Cl = 3HCl&N

:⁝N+NH^3 = 2·N&3·H

4:⁝N&3Cu = 3:⁝N·Cu&:N

:⁝N·K&2·H⁝S = 2⁝S+·K&2·H+:⁝N

:⁝N = ·⁝N&O
:⁝N&:N = ·⁝N&⁝N
2:⁝N&:N = 3·⁝N
2·⁝N = :⁝N&⁝N
3·⁝N = 2:⁝N&:N
2⁝N = ·⁝N&:N

3⁝N = :⁝N&2:N
H&O = ·H
·H&O = :H
H&·N^2 = ·H&N^2
H&·N = ·H&N
2H&:N = 2·H&N

:N&O^2 = ·⁝N
:N&2·N^2 = ·⁝N&N^4
3:N&4·N^2 = ·⁝N&2⁝N&N^5

S&O^2 = :S
:S&⁝N = ⁝S&:N
·H&SFe&⁝S = ⁝S·Fe&HS
HS&O^3 = ·H&:S

2:B+·Na&⁝S = ⁝S·Na&2:B

:⁝P^2&5C = P^2&5·C

C&O^2 = :C
:C·Ca&HCl = HCl+·Ca&:C
:C&C = 2·C
:C·Ca&Fe = ·C&·Ca&·Fe
OC^{2}H^3 = ·H&2HC
⁝C^2 = ·C&:C
HC&O^3 = ·H&:C
H^{2}C&O^4 = 2·H&:C

2HCl&:Mn = HCl·Mn&Cl&·H
2HI&:Mn = HI·Mn&I&·H
6Cl&6·K&5·H = 5(HCl·K)&:⁝Cl·K
3:⁝Cl·K&4⁝S = 2(2⁝S+·K)&·⁝⁝Cl·K &2·⁝Cl

HI&Cl = HCl&I
HI&⁝S = ·H&:S&I

HBr+·Mg&Cl = HCl+·Mg&Br

FCa&·H⁝S = HF&⁝S·Ca

:C·K&Ca = ·K&:C·Ca
⁝S·K&4C = SK&4·C
·H&SK = HS·K
4S&4·K = ⁝S·K&3SK
2*T*+·K&:C·K = 2*T*·K&:C

·H&Fe&I = HI+·Fe
HI·Fe&:C·K = HI·K&:C·Fe
HI·K = ·H&IK

:C·Na&·Ca = ·Na&:C·Ca
2*T*+·K&:C·Na = *T*·K+*T*·Na&:C
⁝S·Na&4C = SNa&4·C
·H&SNa = HS·Na
HCl+·Na = ·H&ClNa
·H&ClNa = HCl+·Na

HCl+NH^3&·Ca = ·H&ClCa&NH^3
NH^3&O^3 = 3·H&N

:C·Ca = ·Ca&:C
:C·Ba = ·Ba&:C
:C·Sr = ·Sr&:C
:C·Mg = ·Mg&:C

⁝S·Ca&4C = SCa&4·C
⁝S·Ba&4C = SBa&4·C
⁝S·Sr&4C = SSr&4·C

⁝S&:C·Ca = ⁝S·Ca&:C
HCl&:C·Ca = HCl·Ca&:C
:⁝N&:C·Ca = :⁝N·Ca&:C
A&:C·Ca = *A*·Ca&:C

:C·K&⁝S·Ca = :C·Ca&⁝S·K
:C·Na&:⁝N·Ba = :C·Ba&:⁝N·Na

⁝S·Na&HCl·Ba = ⁝S·Ba&HCl·Na

2⁝S·Fe = ⁝S&:S&⁝Fe^2
⁝S·Fe&·K = ⁝S·K&·Fe
⁝S·Fe&:C·Na = ⁝S·Na&:C·Fe
HS+NH^3&⁝S·Fe = ⁝S+NH^3&·H&SFe

A·Pb&*Z* = *A*·*Z*&Pb
A·Pb&:C·K = *A*·K&:C·Pb
A·Pb&⁝S·Mg = *A*·Mg&⁝S·Pb
A·Pb&ClNa = *A*·Na&ClPb
A·Pb&IK = *A*·K&IPb
:C·Pb&HS = ·H&SPb&:C

S^{2}Hg&2Fe = 2SFe&Hg
·H&ClHg&·Ca = HCl+·Ca&·Hg
2·H&Cl^{2}Hg&2·Ca = 2HCl·Ca&:Hg
A·K&:⁝N·Hg = *A*·Hg&:⁝N·K
IK&:⁝N·Hg = :⁝N·K&IHg
2IK&Cl^{2}Hg = 2ClK&I^{2}Hg
2⁝S+:Hg&2ClNa = 2⁝S·Na&Cl^{2}Hg
2⁝S+:Hg&Hg &2ClNa } = 2⁝S·Na&2ClHg

:⁝N·Ag&HCl = :⁝N&·H&ClAg
:⁝N·Ag&HCl +·Na } = { :⁝N·Na&·H& ClHg
ClAg&:C·Na = ClNa&Ag&O&:C

TABLE of the Specific Gravity, Symbols, and Equivalent Weights, of the Gases, and their most important Compounds.

Symbols.	Names, Compounds, and their Elements.	Weight.	Bulk of the preceding weight.	Specific gravity.	Number of Equivalents which express the sp. gr.	Composition of the preceding Compounds by weight and by measure, with the symbols of the component parts.						
						Symbol.	Weight.	Measure.		Symbol.	Weight.	Measure.
H	HYDROGEN	1	□	1	1							
O	OXYGEN	8	▭	16	2							
·H	Oxide of Hydrogen	9	□	9	1	O	8	▭	+	H	1	□
N	NITROGEN	14	□	14	1							
·N	Oxide of	22	□	22	1	O	8	▭	+	N	14	□
:N	Binoxide of	30	□□	15	½	O^2	16	⊟	+	N	14	□
⋮N	Hyponitrous Acid	38	?	?	?	O^3	24	⊟▭	+	N	14	□
·⋮N	Nitrous Acid	46	□	46	1	O^4	32	⊞	+	N	14	□
:⋮N	Nitric Acid	54	?	?	?	O^5	40	⊞▭	+	N	14	□
·N^2	Atmospheric Air	36	□□▭	14⅖	⅖	O	8	▭	+	N	28	□□
NH^3	Ammonia	17	□□	8½	½	H^3	3	□□□	+	N	14	□
S	SULPHUR	16	□?	16	1							
:S	Sulphurous Acid	32	□	32	1	O^2	16	□	+	S	16	□
⋮S	Sulphuric Acid	40	?	?	?	O^3	24	⊟▭	+	S	16	□
HS	Sulphureted Hyd.	17	□	17	1	H	1	□	+	S	16	□
HS^2	Bisulphureted Hyd.	33	?	?	?	H	1	□	+	S^2	32	□□
P	PHOSPHORUS	16?	□?									
HP	Hydruret of	17	?	17?	1?	H	1	□	+	P	16	□
C	CARBON	6	□?	6?	1?							
·C	Carbonic Oxide	14	□	14	1	O	8	▭	+	C	6	□
:C	Carbonic Acid	22	□	22	1	O^2	16	⊟	+	C	6	□
C^2H	Bicarburet of Hyd.	13	?	?	?	H	1	□	+	C^2	12	□□
HC	Hydruret of	7	▭	14	2	H	1	□	+	C	6	□
H^2C	Bihydruret of	8	□	8	1	H^2	2	□□	+	C	6	□
OC^2H^3	ALCOHOL	23	□	23	1	2HC	14	⊟	+	·H	9	□
OC^4H^5	Sulphuric Ether	37	□	37	1	4HC	28	⊞	+	·H	9	□
C^2N	Cyanogen	26	□	26	1	N	14	□	+	C^2	12	□□
HC^2N	Hydrocyanic Acid	27	□□	13½	½	H	1	□				
Cl	CHLORINE	36	□	36	1							
·⋮Cl	Peroxide of	68	□□	34	½	O^4	32	□□	+	Cl	36	□
:⋮Cl	Chloric Acid	76	?	?	?	O^5	40	⊞▭	+	Cl	36	□
HCl	Muriatic Acid	37	□□	18½	½	H	1	□	+	Cl	36	□
I	IODINE	124	□	124	1							
:⋮I	Iodic Acid	164	?	?	?	O^5	40	⊞▭	+	I	124	□
HI	Hydriodic Acid	125	□□	62	½	H	1	□	+	I	124	□
Br	BROMINE	78	□	75	1							
HBr	Hydrobromic Acid	79	□□	37½	½	H	1	□	+	Br	75	□

1st column contains the symbols of the compounds.

2d column contains their names.

3d column contains the weight of an equivalent quantity of each.

4th column contains the bulk or measure of the weight referred to in the preceding columns, hydrogen being taken as a standard of comparison. The square figure represents a whole measure, and the others indicate proportional quantities. Thus, in reading the third line, we say, "·H is termed oxide of hydrogen (water), its equivalent weight is 9, which corresponds with one measure.'

5th column shews the specific gravity of the gases; hydrogen being taken as a standard of comparison.

6th column shews the relation between the equivalent number and the specific gravity which the same bulk of gas (one measure) must always be taken.

Lastly, the remaining columns indicate the quantity by weight and by measure of the elements, in the proportion referred to in the preceding columns; the rest of the third line is therefore read in the following manner,—" being composed of oxygen 8, or half a measure; 8 parts of oxygen filling the measure half full, while 1 of hydrogen, which is so much lighter, fills it entirely." Though the equivalent by weight of a compound is found by adding the equivalents of its elements, the corresponding bulk is often less than the bulk of both separately, the particles packing, as it were, in less space when combined.

ABBREVIATIONS.

Temp.........temperature. Fahrenheit's thermometer is always referred to, if no other is particularly mentioned.

Bar............barometer.

□ ....one measure or volume ; the square representing cubic capacity.

▭ ....half a measure or volume.

Parts.........always signify parts by weight, where it is not otherwise stated.

W ....weight.

V ....volume, or measured bulk.

+ ....is translated " combined with," (see p. xii), and signifies to be added to, or mixed with.

× ....multiplied by.

= ....equal to, or composed of.

c. i. ....cubic inches.

gr.............grain.

sp. gr. ......specific gravity.

par. ....paragraph.

Symb. ......symbol.

Synon. ......synonymes.

Eq. ....equivalent.

The Student is recommended to write frequently Symbols and Diagrams, illustrative of the most important cases of chemical action, after he has become familiar with the preceding Tables.

INTRODUCTION.

1. THE properties of matter are Primary and Secondary.

2. The PRIMARY PROPERTIES, EXTENSION, IMPENETRABILITY, INERTIA, and INDESTRUCTIBILITY, are observed in all bodies. Impenetrability signifies that no portion of matter can occupy the same place at the same time with another; Inertia, that it cannot of itself begin to move or assume a quiescent state after being put in motion, or that it has no spontaneous motion.

3. The SECONDARY PROPERTIES of matter include those which are not essential to the existence of matter; they are GENERAL or PARTICULAR.

4. ATTRACTION, or that power by which matter attracts matter, is universal, and is ranked among the General Secondary properties. The Particular properties, as colour, texture, form, &c. serve to distinguish one matter from another.

5. No property of matter is more conspicuous than ATTRACTION. Operating between masses of matter, it is termed GRAVITATION; COHESION, or the ATTRACTION OF AGGREGATION, when it is exerted between particles of the same kind; CHEMICAL ATTRACTION or AFFINITY *, when it operates between particles of different kinds of matter. The MAGNETIC AND ELECTRIC ATTRACTIONS are induced by peculiar arrangements.

6. The weight of bodies is estimated by the force with which they are attracted to the earth. The weight of equal bulks of different substances compared with some standard of comparison, is termed their DENSITY or SPECIFIC GRAVITY. The annexed table shews the relative weights of equal bulks (the Specific Gravity) of five different substances, hydrogen being taken as a standard of comparison, and air considered 828 times lighter than water.

Hydrogen, the lightest body known,	1.
Atmospheric Air, ...	14.4
Oxygen, ...	16.
Water, ...	11923.
Platinum, the heaviest body known,	256345.

7. Atmospheric air is generally taken as a standard of comparison in estimating the density of gases, and water for liquids and solids. As heat influences the bulk of all bodies, experiments on specific gravity must be made at a given temperature; 60 of Fahrenheit is usually taken as a standard temperature in this country. In stating the specific gravity of gases, the height of the barometer, and their condition with respect to moisture, must also be particularly noted, as these influence the result.

8. The air presses upon the surface of the globe, in the same manner as the water upon the earth beneath it. It presses with the same force as a weight of 15 lb. laid upon every square inch of surface. A column of mercury 30 inches in height,

* This term was introduced on the supposition, not now entertained, that those substances only attract each other chemically which resemble each other.

or of water nearly 34 feet high, presses with the same force; and, accordingly, a long tube closed at one end, and filled with mercury, still remains full to the height of 30 inches, as in the common barometer-tube, when inverted in mercury; the rest of this liquid escaping, the air and the mercury balance each other; the space between the upper portion of the mercury and the tube contains nothing; it is called the Torricellian vacuum, from Torricelli. In making this experiment, the height of the mercury is not always the same, varying from about 1 to 3 inches with the pressure of the atmosphere.

9. The DIVISIBILITY even of very small portions of matter, as a grain, into many millions of parts, has long been familiar. Matter has been regarded as infinitely divisible, but this opinion is no longer entertained. From recent discoveries in Chemistry (see Atomic Theory), it is considered extremely probable that all masses of matter are composed of particles or ATOMS, totally indivisible by any power to which they may be subjected in the ordinary course of nature, but having all the essential properties of matter as truly as any masses which they may compose. They may be divisible in a mathematical point of view, but they are physically indivisible. The extreme minuteness of these ultimate particles, which may far exceed even the minutest portions of matter that have been observed, has hitherto prevented many interesting circumstances connected with their history from being determined.

10. POROSITY is a property of all masses of matter, even of those that appear most dense. The pores are the spaces which exist between the particles of which the matter is composed, and being often filled with air, water, or other substances, the action of the included material sometimes affects much the result of numerous chemical processes.

11. COMPRESSIBILITY by mechanical force is another property of masses of matter, which proves that their particles are not in the nearest possible contact. Gases are much more compressible than solids or liquids. Those bodies that recover more or less their former bulk, when the compressing cause is removed, are said to be ELASTIC. All liquids and gases recover their original bulk when freed from pressure.

12. Air, and all gasses at the surface of the earth, are much compressed by the pressure of the superincumbent air. When they are relieved from this pressure, as in working the common air-pump, they expand to a great degree, so that it is easy to procure in this manner a vessel containing only 1–1000dth part of the air or gas originally present.

13. POLARITY, or a disposition in the particles of matter to move in a regular and determinate manner, and not in a confused mass, when affected by other agents, is a property of all kinds of matter.

14. Lastly, Matter is powerfully affected by the action of HEAT, LIGHT, ELECTRICITY, and MAGNETISM. These forces sometimes promote attraction; on other occasions they establish a repulsion between the particles of matter, and cause them to separate,—they communicate a power of REPULSION which causes a separation of the parts affected. By *Attraction* and *Repulsion* all changes in matter are regulated.

15. Other secondary properties of matter, as Hardness, Transparency, &c. it would be superfluous to enumerate here. *

* The reader is requested to examine the contents generally before proceeding farther, that he may become familiar with the arrangement adopted in this work.

PART I.

GENERAL PRINCIPLES OF CHEMISTRY.

I. CHEMICAL ACTION.

16. All bodies in nature, in a chemical point of view, are arranged into ELEMENTS and COMPOUNDS. All those substances which cannot be resolved into other kinds of matter, are considered simple and termed ELEMENTS; thus lead is called an element, because nothing but lead can be obtained from it. COMPOUNDS contain different elements, however simple and uniform they may appear to the eye, as the sulphuret of lead, the galena of mineralogists, from which both lead and sulphur can be procured.

17. The smallest particle of an element or compound is termed an INTEGRANT PARTICLE; the particles of the different elements of compounds are called their COMPONENT or CONSTITUENT PARTICLES.

18. When several compounds are associated together, they are frequently termed PROXIMATE PRINCIPLES. ULTIMATE PRINCIPLES are merely the elements of which these proximate principles are composed.

19. CHEMICAL ACTION consists in Combination or Decomposition, and often both occur at the same moment. COMBINATION, or SYNTHESIS, is the union of particles of different kinds of matter: it is very different from MIXTURE, the particles cannot be separated again by any mechanical means.—DECOMPOSITION, called also ANALYSIS, is the separation of particles previously united.

20. Before chemical action can take place, the particles between which it is exerted must be brought by mixture into the nearest possible contact.

21. All bodies do not act chemically on each other. This is not attributed to a total want of chemical attraction between them, but rather to the attraction being rendered inefficient by other causes.

22. Compounds can often combine with elements, and also with each other; and complicated combinations are often subverted without reducing any of the ingredients to the elementary form, proximate principles alone appearing.

CHAP. I. PHENOMENA OF CHEMICAL ACTION.

23. The phenomena attending chemical action are extremely various, as the evolution of heat and light, a change in the form, density, tenacity, power of attraction for different bodies, and also in relation to heat, light, and electricity. Occasionally, however, no phenomena are apparent. The change of properties accompanying chemical action is often so great that the eye cannot see the most distant resemblance in the resulting product to any of the ingredients employed.

24. NEUTRALIZATION is a term employed to denote the complete loss of characteristic properties that frequently attends combination in particular proportions; the compound formed is said to be NEUTRAL, and the one ingredient is said to be NEUTRALIZED or SATURATED by the other. The latter term is also employed in a different sense, to indicate the largest quantity of one body that can combine

with a given weight of another, as when water is said to be SATURATED with salt, or has taken up as large a quantity of it as it can dissolve.

25. The change of properties attending chemical action is often not so very marked where the attraction is feeble, or the combining substances resemble each other much, as the different metals.

26. The density is generally increased by combination, and diminished by decomposition. In the combination of gases, when condensation occurs, it bears a very simple ratio to the bulk of the combining ingredients. Kirwan has remarked, that, in the combination of some liquids, the full amount of condensation does not take place till many hours after the compound shall have been formed.

27. PRECIPITATION is the falling down of a solid from a liquid or gas with which it may previously have been combined. EFFERVESCENCE is the rapid disengagement of a gas from a fluid. SOLUTION indicates the combination of a solid with a liquid or gas, in which it dissolves in the same manner as sugar in water.

28. Chemical action is said to be DIRECT, or PRIMARY, where it takes place solely between the substances used, while they are still in the condition in which they were mixed. It is termed INDIRECT or SECONDARY, where part of the products formed at first begin to react on the materials originally employed.

29. To illustrate complex cases of chemical action, diagrams were introduced by Dr Black, and have been much employed by succeeding chemists. The following example illustrates a series of new diagrams which I have constructed, adapted to the present state of chemical science. The names of substances that react on each other, and numbers representing the proportions, are placed on the left, one above the other. When any of these are compounds which are decomposed, brackets proceeding from them enclose the names of their elements. Lines traced from these numbers represent the quantities of each, proceeding from the left to the right. Those that proceed from the different elements or compounds that combine together terminate in a common point, and opposite these, towards the right, are placed the names of the products and their quantities. The annexed figure represents the decomposition that takes place during the preparation of ammonia from the muriate of ammonia by means of lime.

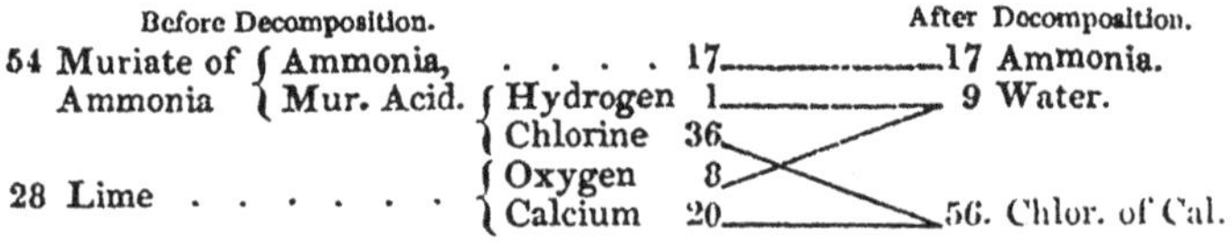

30. The character of the line expresses the different conditions in which the products of any reaction are obtained. When they are gaseous, the lines are dotted, as in those pointing to ammonia. When they are liquid or retained in solution, lines composed of many smaller lines are used, as in those pointing to water. And when any substance is obtained in the solid form, a plain line is drawn. Any variety of form may be given to the diagram according to the effect which has to be represented.

CHAP. II. FORCE OF AFFINITY.

31. The force with which chemical action is exerted is extremely great, the particles of the most cohesive metals being torn from each other with facility; but there is no method by which its absolute force can be ascertained. Nor can the relative power with which different substances act on each other be precisely measured, as this is so various under different circumstances; thus lime attracts carbonic acid quickly at ordinary temperatures, but when heated, they do not combine.

32. The different examples where two ingredients in combination are separated by a third substance attracting one, have been termed by Bergman cases of SIMPLE ELECTIVE AFFINITY. When two compounds decompose each other, two new compounds being formed by an exchange of elements, it is said to be a case of DOUBLE ELECTIVE AFFINITY. It is also called DOUBLE DECOMPOSITION, or COMPLEX AFFINITY.

33. QUIESCENT AFFINITIES signify those that tend to maintain the elements of a compound in their present state: by DIVELLENT AFFINITIES are understood those that tend to arrange the particles of a compound in a new form, producing decomposition. Decomposition necessarily ensues on mixing different compounds, if the total sum of the divellent be stronger than that of the quiescent affinities.

34. Bergman considered combination and decomposition to depend solely upon the strength of affinity. Berthollet pointed out that affinity is modified by numerous circumstances.

35. If the decomposition of a compound cannot be effected by the operation of a single agent, two or more can be brought into play, as in double decomposition, and these will frequently at once effect the decomposition required.

CHAP. III. CAUSES WHICH MODIFY AFFINITY.

36. The principal circumstances modifying chemical action are the relative quantities of materials used, cohesion, elasticity, insolubility, specific gravity, state of combination, light, heat, and electricity. The influence of the last three will be seen on referring to light, &c.

SECT. I.—*Quantity of Matter.*

37. The larger the quantity of any ingredient combined with another, the more easily is a portion of it separated.

38. The less the quantity of any substance united with another, the more readily does an additional quantity enter into combination.

39. The larger the quantity of any chemical agent brought into play, if not excessive, the more speedily will its action be effected. But very different products are occasionally obtained, according to the proportions used.

40. The chemical agency of all bodies must necessarily be in the ratio of the *strength of affinity* exerted at the moment they are used, however modified that may be, and the quantity in which they can be brought to act effectually on those ingredients of a compound which they can separate. But the quantity of a decomposing agent which can be mingled with a compound to be decomposed is unlimited; whereas the portion that can be brought into effective action may be comparatively small at times, and incapable of producing any effect. Berthollet, from not attending to this distinction, has frequently advanced propositions, in his valuable remarks on affinity, which are quite untenable.

SECT. II.—*Cohesion.*

41. The attraction of aggregation, or cohesion, as it binds the particles of bodies together, opposes chemical action, or any force that may tend to separate them. Cohesion is occasionally so great, that no chemical action ensues between numerous substances till it has been diminished.

42. The greatest degree of cohesion occurs in solids. Liquids have very little cohesion, and in gases it is not perceptible; their particles are believed to repel each other.

43. Cohesion may be diminished *mechanically* by grinding, rasping, pulverizing, &c.; *chemically*, by the action of solvents, and also by the operation of heat. Solids are often procured with great facility, in a very minute state of division, by solution and subsequent precipitation.

44. Cohesion is often alluded to by Berthollet, and some modern authors have adopted his views, as a power which at times determines the union of different ingredients to the exclusion of a third, though the cohesive power spoken of cannot actually exist till the compound whose formation it is to promote has been formed. It is an effect of the combination, not a cause.

SECT. III.—*State of Combination—Elasticity—Insolubility, Specific Gravity, &c.*

45. STATE OF COMBINATION.—A body in combination with another may not combine with a third substance, being retained by a stronger affinity; or the reverse may take place as it may be brought by combination into such a form that it now unites readily where it could not previously enter into combination.

46. The tendency of many bodies to enter into combination is often promoted by presenting to them a third substance, which exerts a strong attraction to the compound they form, all the substances brought into action affecting each other at the same moment; but the combination having once been established, the third substance may then be withdrawn. Cases such as this are termed examples of DISPOSING AFFINITY.

47. Bodies coming into contact with each other at the instant they may be leaving some previous combination, are peculiarly prone to combine with each other. They are then said to be in their NASCENT STATE.

48. ELASTICITY.—Bodies in the gaseous state are not in general so prone to enter into combination as when their elasticity has been reduced by cold, mechanical pressure, or chemical combination.

49. When many gases are associated together in a condensed form in any combination, the compound formed is in general decomposed with facility by heat; the volatile ingredients having their elasticity always increased by heat, their expulsion is necessarily facilitated by an elevation of temperature.

50. In the same manner, caloric, by increasing elasticity, may be stated generally to promote the separation of a more volatile from a less volatile ingredient.

51. But heat often causes gases to combine, though its first effect must be to add to their elasticity, as in the explosion of various gaseous mixtures by flame.

52. SPECIFIC GRAVITY.—A difference of specific gravity opposes combination, by preventing that intimate mixture which is so favourable to chemical action.

53. The INSOLUBILITY of bodies in different fluids prevents their cohesion being so much reduced in various cases, as might be necessary to promote a rapid chemical action. Any insoluble matter, when precipitated, being removed to a certain extent from the sphere of action, will not be so apt to induce fresh changes, as if it had remained in solution. A liquid decomposing a solid, and forming an insoluble compound with one of the ingredients, often produces a crust of insoluble matter round every little mass of the solid, and thus at times completely arrests the progress of the decomposition.

CHAP. IV. PROPORTIONS IN WHICH BODIES COMBINE—ATOMIC THEORY.

54. I. A few substances can unite apparently in any proportion in which they may be mingled, as alcohol and water.

55. II. Others combine in any proportion till a certain quantity of one ingredient has been added, and then no farther combination can ensue; any quantity of salt, for example, however great or small, can be dissolved in a given weight of water, provided it does not exceed a definite quantity; any portion beyond this is not affected by the water.

56. III. Some substances can combine only in one, or in a few fixed propor-

tions; the laws which have been discovered with respect to these form one of the most important parts of Chemistry; they are the subject of the following remarks.

57. *There is a relation between the proportions in which all bodies combine with each other. The numbers expressing the proportions in which any two bodies combine with a given weight of a third substance, express also the proportions in which these bodies are disposed to combine with each other.* Thus 1 of hydrogen combines with 6 of carbon, and with 8 of oxygen; but 6 of carbon combine with 8 of oxygen.

58. The terms EQUIVALENT or CHEMICAL EQUIVALENT, PROPORTION or COMBINING PROPORTION, PRIME, ATOM, COMBINING RATIO, are in common use to indicate these proportions referred to a standard of comparison. Hydrogen is generally taken in this country as the standard, the number assigned to it is 1, 10, or 100; oxygen has been preferred by some chemists.

59. The equivalents of bodies are ascertained by observing the proportion in which they combine with the standard of comparison, or with an equivalent weight of any substance, and the ratio in which it increases in each successive combination, when it can form different compounds with the same substance. The equivalents of compounds are composed of the added equivalents of their elements; thus the equivalent of water is 9 = oxygen 8 + hydrogen 1.

60. The tables of equivalents should be studied by the beginner till he is quite familiar with them; they communicate much information. Dr Wollaston constructed a SLIDING SCALE for facilitating calculations with respect to the proportions of materials in different combinations, and the quantities required in experimenting. It is much employed by practical chemists; the method of using it is detailed in the explanation which accompanies it.

61. *When a body combines with another in more than one definite proportion, the quantity of one of them in the different combinations is found to be exactly double, triple, or some multiple, by a whole number, of the smallest proportion in which it enters into combination with the other substance.* Thus 14 of nitrogen can combine with 8, 16, 24, 32, or 40, of oxygen, but with no intermediate proportion.

62. When the bulk or volume of gases is compared with their equivalent weights, Gay Lussac showed that there is a very simple ratio between the volumes of all the gases. The equivalents, for example, of hydrogen, nitrogen, and chlorine, are, by weight, 1, 14, and 36; but when these different weights are measured, they are all found to occupy the same space. The table of equivalents by weight and by volume shows the correspondence between the equivalents of the most important gases by weight and by volume, illustrating what is usually referred to under the title of the THEORY OF VOLUMES. Hydrogen is taken as a standard of comparison.

63. It is to Dalton that science is indebted for the ATOMIC THEORY, which points out the law that regulates all these various combinations. Wenzel, Richter, and Bergman, had noticed several of the facts which are included in his more general propositions, and Higgins had entertained similar opinions, though they were not announced in such precise terms, nor in such a form, as to make them generally known. The labours of Wollaston, Thomson, Berzelius, and Gay Lussac, have contributed much to the extension and illustration of his views.

64. If we admit with Dalton, that masses of matter are composed of ATOMS or INDIVISIBLE PARTICLES, differing in weight in different kinds of matter, and that chemical action takes place between these, then a relation between the proportions of combination by weight must necessarily exist. For if bodies combine particle to particle, then the whole weight of the masses entering into combination will be as the weights of their particles, and the relative weights of the masses must accordingly express the relative weights of the particles. And, thus, when 1 of hydrogen combines with 6 of carbon and 8 of oxygen, it is inferred that 6 of carbon

will combine with 8 of oxygen, for these numbers express the weight of their particles, and neither of them can be broken down.

65. Where the same substances combine in different proportions, the quantity of one being fixed, the increase in the other must be a multiple of the smallest quantity that can enter into combination. Thus, if 8 express the combining proportion or weight of a particle of oxygen, the smallest increase to this that can be made must be 8 more, as the particle is indivisible, and hence 16 is the quantity of oxygen found in the second combination, 24 in the third, and so on.

66. Many chemists have entertained the idea, that the equivalents of all bodies are multiples of hydrogen by a whole number, in consequence of some circumstances pointed out by Dr Prout. The result of numerous analyses do not justify us in admitting this to be the case, but several are inclined to attribute this want of correspondence more to unavoidable errors in manipulation, than to any inaccuracy in the opinion itself. Future experiments alone can determine this question.

67. Those cases where bodies combine in apparently unlimited proportions, have not hitherto been so particularly examined. The compounds produced are perhaps formed by the union of a few definite compounds. Others, again, have suggested the idea, that combination in unlimited proportions between different bodies depends upon some peculiarity or resemblance in the form of their particles.

68. A new field of inquiry has presented itself of late years in the discovery of ISOMERIC COMPOUNDS, in which different elements are associated together in the same proportions as in combinations already known, the difference in properties being caused by a difference in the arrangements of the particles; the term is derived from two Greek words,—ισος, equal, μερος, a part.

69. The prefixture PARA is often applied to distinguish one variety of such combinations from others, as in tartaric and paratartaric acids.

70. When the ratio of the elements is the same in different compounds, though the total number of each is greater in one compound than in the others, they are termed POLYMERIC compounds.

71. METAMERIC COMPOUNDS are those where the ultimate elements are the same as in other well known combinations, but considered to be arranged in a very different way; thus, where oxygen, hydrogen, sulphur, and a metal, are associated together, they may be considered to be combined in the form of sulphureted hydrogen and a metallic oxide, or of water (consisting of oxygen and hydrogen) and a metallic sulphuret.

72. In stating the equivalent numbers of compounds, the primary numbers adopted ought carefully to be kept in view, and no deviation from them permitted; if this were carefully attended to, beginners would not be so often perplexed by statements apparently contradictory of the very principle on which the atomic theory is founded. Thus, one of the oxides of manganese is often represented as a compound of 28 manganese and 12 oxygen, instead of 56 manganese and 24 oxygen. The ratio indeed is the same in both, but in the first case the composition is not expressed in the terms of the system on which the equivalent numbers are founded. So many deviations from the proper system have been made, that even those who are opposed to them are occasionally compelled, for the sake of uniformity, to adopt the same expression. The expression of *half atoms*, and similar contradictory terms, have resulted from this practice. The student who attends to the ratio of the weight in such cases, and the proper equivalent numbers, will be able to translate them correctly.

CHAP. V. CHEMICAL NOMENCLATURE.

73. Chemistry has extended so much of late years, and so many anomalies have crept into its nomenclature, that the whole subject demands revision, and to be adapted to the present state of science.

74. The names of elements are usually derived from some of their more characteristic properties.

75. The terms BINARY, TERNARY, QUATERNARY COMPOUNDS, &c. refer in this work to the number of elements or proximate principles in a compound, not to the number of atoms.

76. In binary compounds of oxygen, chlorine, iodine, bromine, and fluorine, which are not acid, the name of the compound terminates in IDE, as in oxide of zinc, chloride of mercury, and iodide of lead.

77. In binary compounds of all other substances, the name terminates in URET, as in hydruret of carbon, sulphuret of iron, potassiuret of antimony.

78. The number of atoms in the first element or proximate principle mentioned in any compound, is usually indicated by a Latin numeral, as BIS, TER, QUATER, &c., and the number of atoms in the second by a Greek numeral, as DIS, TRIS, TETRAKIS. No prefixture is made when the compound consists of 1 atom of each ingredient.

79. To the rules in this paragraph there are many exceptions. Thus protoxide and deutoxide are familiarly used as substitutes for oxide and binoxide.

80. DEUTOXIDE is frequently used to signify a compound of 3 atoms of oxygen with 2 of metal, as in the deutoxide of manganese, and deutoxide of lead.

81. The prefixture PER indicates an oxide containing the largest quantity of oxygen that can exist in any oxide formed of the same materials.

82. OXACIDS are acids containing oxygen, and the relative number of atoms of oxygen, where different acids are produced by the same element with this substance, are indicated by the prefixtures or the termination of the name.

83. IC generally indicates the compound with most oxygen, as in sulphuric acid.

84. OUS indicates a smaller quantity of oxygen, as in sulphurous acid.

85. HYPO indicates a smaller quantity of oxygen than is found in compounds to whose name it is prefixed, as in hyposulphuric acid and hyposulphurous acid.

86. SUB is occasionally employed to denote an intermediate degree of oxidation, as in subsulphurous acid, which contains less than sulphurous, but more than hyposulphurous acid, according to Dr Thomson.

87. PER is used where acids have been discovered containing still more oxygen than those whose names terminate in ic, as perchloric acid.

88. HYPER where a still larger quantity is observed.

89. HYDRACIDS are acids containing hydrogen; their names are also made to terminate in IC, as hydriodic acid, composed of hydrogen and iodine.

90. Aqueous is a term now coming into very general use in designating definite combinations with water; the word Hydrate has long been employed for the same purpose: a prefixture is made when there is more than one atom, as in binaqueous, terhydrate.

91. ALCOATES are definite compounds of alcohol and various saline substances; they were discovered by Mr Graham.

92. SESQUI is a prefixture applied to compounds where two atoms of one substance are united with three of another, as in the sesqui-carbonate of ammonia, where two of ammonia are united with three atoms of carbonic acid. All such combinations may be considered as constituted of two compounds, in one of which the ingredients are combined atom to atom, and in the other in the proportion of

two to one; in the example quoted, the sesqui-carbonate of ammonia may be regarded as a compound of the carbonate and bicarbonate of ammonia. The term HEMI has been proposed by Mr Lunn as a prefixture in similar cases.

93. The term SALT is applied to a very extensive range of compounds, where acids are combined with oxides or other compounds having similar properties, as ammonia. The oxide, ammonia, or other substance united with the acid, is called a BASE, or SALIFIABLE BASE.

94. The NAMES OF SALTS are composed of the names of the acid and base.

95. ATE is the terminating syllable given in such cases to the name of an acid ending in IC, as in sulphate of potassa, composed of sulphuric acid and the base potassa.

96. ITE is the terminating syllable when the name of the acid ends in OUS, as in sulphite of potassa, which consists of sulphurous acid and potassa.

97. A SUPERSALT indicates that the properties of the acid predominate, without reference to the precise composition, as in the supertartrate of potassa.

98. A SUBSALT indicates that the qualities of the base predominate, as in the sub-borate of soda.

99. A NEUTRAL SALT expresses that the compound has none of the characteristic properties either of the acid or base, as in the sulphate of potassa; the term neutral is often omitted.

100. The three last prefixtures being used only in a general sense, other terms are necessary to express the precise constitution of saline combinations.

101. No PREFIXTURE is in general made when the salt is composed of 1 atom of acid and 1 of base, the latter being composed of 1 atom of metal and 1 of oxygen, if it be an oxidated metal. The term proto, however, is occasionally prefixed to express this more specifically, as in protosulphate of iron for sulphate of iron.

102. In salts of the common metals, the name of the metal is substituted for the name of the base, where there is but one atom of oxygen united with one of metal, as in sulphate of iron, which is more correctly termed sulphate of the oxide of iron.

103. BIS, TER, QUATER, &c. are prefixed where 2, 3, or 4 atoms of acid are combined with 1 of base, as in the bisulphate of soda, and quadrisulphate of potassa.

104. DIS, TRIS, TETRAKIS, &c. are prefixed where 1 atom of acid is united with 2, 3, 4, or more atoms of base, as in the dichromate of lead, and trisnitrate of bismuth.

105. A DOUBLE PREFIXTURE is necessary, when there is more than 1 atom of oxygen in the base, as well as an unequal number of atoms of acid and base, as in the bipersulphate of mercury, where *bi* expresses 2 atoms of acid, and *per* that the mercury is in the form of a peroxide.

106. DOUBLE SALTS are compounds where one salt is combined with another, as in the tartrate of potassa and soda, regarded as a compound of tartrate of potassa and tartrate of soda. The term Triple Salt is also applied to such combinations, the name being derived from the three different ingredients present.

107. HALOID SALTS differ much in their composition from the preceding substances; the term is applied to compounds of the metals with bodies such as chlorine, iodine, bromine, fluorine, and cyanogen; but not to compounds containing oxygen and sulphur. The term is derived from αλς, salt, and ειδος, form, as they are similar in constitution to common salt.

108. Haloid salts can combine with different acids, oxides, common salts, and also with each other.

109. SULPHO SALTS include a series of compounds which have been particularly attended to only of late years; they consist of compounds, both of which contain sulphur. Compounds of hydrogen, carbon, or metals with sulphur, when

combined with other compounds formed of metals and sulphur, constitute this group of salts. In the common salts, a large proportion of which contain oxygen, this element is found both in the acid and in the base: thus, in the phosphate of soda, it is associated with phosphorus in phosphoric acid, and with sodium in the soda; and hence such combinations are termed **Oxy-Salts.** In the **Sulpho Salts**, sulphur plays the same part, forming with one element an acid, and with the other a salifiable base.

110. Ammoniurets are compounds containing ammonia and salifiable bases, or other substances not acid.

111. The prefixture *ammoniaco*, or *ammonia*, is frequently put before salts, where this substance has been added in sufficient quantity to combine both with the acid and the base.

112. These remarks will enable the beginner to see the leading features of the present system of nomenclature. Lavoisier, Fourcroy, Guyton-Morveau, and Berthollet, laid the foundation of a system of nomenclature on scientific principles. The observations of Berzelius, Dalton, and Thomson, have done much to extend it according to the progress of the science; and in the Encyclopædia Metropolitana there is an interesting paper on this subject by Mr Lunn. A more rigid observance of this system would save much trouble and perplexity both to beginners and professional chemists.

113. Chemical symbols are now much employed, and with very great advantage; but here also a difference of opinion as to the most advantageous system has been productive of much inconvenience. While there is no general standard of reference in this country, I shall continue to employ that explained in the Table of Symbols (See Tables).

CHAP. VI. CRYSTALLIZATION.

114. In numerous chemical operations, some of the products obtained are separated in masses of regular symmetrical forms. These have been denominated *crystals*, and the process by which they are procured is termed crystallization.

115. The formation of crystals depends upon the cohesive attraction of the particles being exerted in a slow and equal manner, when they have perfect liberty of motion. To crystallize bodies, therefore, their cohesion, when they are solid, must be diminished, in the first place, either by heat or by solution.

116. When fluidity is communicated by the action of a solvent, a hot saturated solution is generally made, and crystals are deposited as it cools. The dissolving power of fluids increasing for the most part with the temperature, that part solely of the substance which is dissolved in consequence of the high temperature, crystallizes as the liquid cools. The rest remains in solution.

117. If the remaining fluid be removed and concentrated by boiling, additional crystals may be procured; and this may be repeated with the successive drainings till they are all crystallized. These processes cannot always be resorted to, as there are several substances which are more soluble in cold than in boiling water.

118. When a *pellicle* or crust forms on the surface of a solution while it is concentrated, this shews that it is already saturated; the formation of the pellicle depends upon part of the water being expelled which was necessary to retain the matter of the pellicle in solution.

119. When the solvent is removed by rapid heat, the solid matter falling down in the boiling fluid, as in the preparation of common salt, the process is termed salting.

120. When the solvent is removed slowly by spontaneous evaporation, larger

and more regular crystals are procured than when crystallization is effected by a sudden reduction of temperature.

121. The stronger the solution, the more rapid is the crystallization, and the smaller the crystals.

122. When several crystallizable substances having little attraction for each other are present in the same solution, that which is largest in quantity and least soluble crystallizes first. It does not all separate at once; when a portion has fallen, the least soluble salt next in quantity begins to appear, and layers of different salts are thus often deposited successively from the same liquid. This is termed ALTERNATE CRYSTALLIZATION.

123. When water (or other fluids) are employed to dissolve solids for crystallization, a portion of the solvent is often found chemically associated with the crystal. It is termed WATER of CRYSTALLIZATION.

124. Crystallized salts containing water, that melt as they are heated, are said to undergo the WATERY FUSION. The DRY FUSION is the liquefaction produced by heat after the water has been expelled.

125. DELIQUESCENCE is the liquefaction of a salt or other substance by the action of moisture attracted from the air. EFFLORESCENCE indicates the production of a powdery-looking matter produced by the action of a dry atmosphere removing the moisture from a salt.

126. DECREPITATION denotes the separation or splitting of the crystals, accompanied by a crackling noise, and produced by the action of heat.

127. AIR has considerable influence upon crystallization, promoting it by removing a portion of the fluid holding the crystalline matter in solution. A hot saturated solution of sulphate of soda does not crystallize on cooling, if the air be excluded; on admitting the air, crystallization commences. Professor Graham found that the greater the attraction for water in the air or gas admitted, the more certain the effect.

128. A NUCLEUS, or crystal of the same kind with the crystalline matter of any solution, also promotes crystallization. Any solid matter is often used for the same purpose in manufactories, as wooden spars, iron hoops, and threads suspended in the crystallizing fluid: they afford numerous points of contact where the crystallization may commence.

129. It is a modern discovery, that some solids, by long continued heat, can undergo a change in the form of their crystals without being rendered fluid.

130. All bodies tend to crystallize in a peculiar form, so that the form of the crystal is often regarded as one of the best indications of its purity. It cannot, however, in all cases, be implicitly relied on.

131. Some bodies crystallize in the same form, though totally different from each other as chemical agents. They are said to be ISOMORPHOUS, and this new branch of inquiry, which was discovered by Mitscherlich, is termed ISOMORPHISM, from ισος, equal, and μορφη, form. The compounds of isomorphous bodies with similar substances form similar groups of crystals. Where the form of bodies is very similar, though they may not entirely coincide, they are termed PLESIOMORPHOUS, from πλησιος, near.

132. Isomorphous bodies tend to crystallize together, and a portion of one may replace a portion of another, without affecting the form of the mass in which it occurs.

133. In every crystal, Hauy has pointed out that there is a PRIMITIVE FORM or nucleus, which may be procured by splitting or cleaving it in the direction of the layers or laminæ of the crystal. The various forms produced in crystals by combination of the primitive form, are termed SECONDARY or DERIVATIVE forms

134. It was considered, till very lately, that the primitive form or nucleus is always the same in some substances, however various the external form might be, or by whatever process it might have been procured. Later observations have thrown doubt upon this, and some substances, as sulphur, have been remarked to afford two varieties of crystals, when treated in different ways, whose primary forms are incompatible with each other. Such bodies are said to be DIMORPHOUS, from δις, twice, and μορφη, form.*

II. CALORIC.

135. CALORIC is the name given to the cause of all the phenomena attributed to heat and cold. Its precise nature is unknown. Three opinions have been brought forward in regard to it: 1st, Bacon considered that heat consists in a peculiar motion or tremor of the particles of matter; Boyle and Newton adopted his views. 2dly, Many regard it as produced by vibrations in an elastic fluid universally diffused. 3dly, It has been supposed to depend upon a peculiar fluid, invisible, imponderable, and extremely subtle, its particles repelling each other, flying off when largely accumulated, and attracting other kinds of matter, which they penetrate with facility. The second opinion has many advocates at the present time; but the language of the last is more generally adopted, from the simplicity with which the phenomena of heat can be described according to this hypothesis.

136. The properties of caloric are known only by its effects on material objects. All known bodies contain caloric; none can be procured without it, and were caloric utterly withdrawn from them, they might present themselves in some new form, at present altogether unknown to us. It must be kept in mind that every thing in nature is influenced in its appearance and other properties by the heat associated with it.

137. The term ABSOLUTE CALORIC is employed to denote the total amount of heat in bodies. No method is known by which this can be ascertained.

138. The temperature of a body indicates its condition with respect to caloric, referred to a standard of comparison; its temperature in general rises when caloric is communicated to it, and falls when it is withdrawn. The quantity in each continually fluctuates according as it may be affected by the rays of the sun, and numerous natural and artificial operations going on in the globe itself; but wherever any inequality occurs, then the caloric passes from those bodies in which it is accumulated to those in which it is comparatively deficient, tending to establish a universal equilibrium of temperature. Were no disturbing causes continually in operation, the whole globe would at last attain the same temperature. The sensation produced by cold, and all other effects referred to this source, depend upon the abstraction of caloric.

139. The influence of this great power is conspicuous in the change of the seasons, in the alternations of day and night, in its power over the animal, vegetable, and mineral kingdoms, and indeed in every operation, natural or artificial, carried on in the globe.

CHAP. I. EFFECTS OF CALORIC.

140. The most important effects of caloric are,—Expansion, Liquefaction, Vaporization and Evaporation, Ignition, and its effects on chemical action.

* For information with respect to the forms of crystals, and other circumstances connected with this subject, the reader is referred to works on *Crystallography*, which has now become so extended as to form a distinct science.

SECT. I. *Expansion.*

141. Caloric communicated to matter almost invariably expands it. The precise mode in which it produces this effect is unknown; it is believed to insinuate itself between the particles of the expanded bodies, separating them more or less from each other. In general, the less the cohesion the greater the expansion caused by a given increase of temperature; thus, gases expand most, then liquids, and solids least of all. Air expands 3–8ths of its volume, alcohol 1–9th, mercury 1–55th, hammered iron 1–273d, and flint-glass 1–416th, when heated from 32 to 212, or 180 degrees. But this does not apply to individual solids compared with each other, nor to liquids. In solids and liquids the amount of expansion increases in a greater ratio than the temperature, though many of these expand very uniformly through a considerable range of temperature at a distance from their fusing or boiling points.

142. The expansion of solids and liquids takes place with great force. Iron plates and bars are often twisted or broken where no allowance has been made for expansion, and the strongest buildings have been injured from the same cause. In constructing furnaces, and in all cases where metals may be subjected to considerable variations of temperature, this requires to be particularly attended to.

143. The expansion of bodies by caloric continues only so long as the caloric is retained. By a reduction of temperature the original volume is restored.

144. Hence, as bodies contain caloric at natural temperatures, by withdrawing more or less from them, their volume may be more or less diminished.

145. A body exerting great force during its expansion by heat, contracts with great power as it cools. Iron bars, while expanded by heat, having been fixed between the walls of a large building in Paris which had inclined outwards, restored them again to the perpendicular by the force exerted during their contraction as they cooled.

146. Expansion is generally accompanied by a diminution of cohesion.

147. Caloric having no sensible weight, a change of specific gravity attends expansion produced by heat; as the volume increases, the density diminishes, the same weight being now expanded through a larger space. All measures accordingly vary with the temperature, which ought to be particularly noted. A measured ounce of water, alcohol, or ether, at 50°, contains more of these fluids than an ounce by measure taken at the temperature of 100°.

148. Expansion by heat in solids.—Different solids are not expanded equally by the same addition of heat. Each has an expansion peculiar to itself; when heated from 32° to 212°, a rod of lead is elongated 1–351th part; of silver, 1–524th; of gold, 1–602d; of forged iron, 1–819th; of platinum 1–1167th; English flint-glass, 1–1248th part.

149. Expansion of liquids.—These, like solids, have each a degree of expansion peculiar to themselves. In rising in temperature from 32° to 212°, alcohol expands 1–9th of its volume, oil of turpentine 1–14th; water 1–22d; and mercury 1–55th. The expansion of liquids by caloric is familiarly illustrated by heating the common thermometer, or any liquid in a flask. The expansion of any liquid in a glass, or other vessel, always appears less than the actual expansion, the containing vessel being at the same time increased in its capacity.

150. The greater ratio in the increase of expansion as the temperature is augmented, is very marked in several fluids. If the expansion of mercury from 32° to 122° (90 degrees) be stated as 14, the expansion from 122° to 212° (90 degrees) is 15. The expansion of water from 32° to 122° being 4.7, its expansion from 122° to 212° is 15.

151. Expansion of Gases.—Equal increments of temperature produce the

same amount of expansion in all gases, as proved by Dalton and Gay Lussac. For every degree of Fahrenheit's thermometer, gases expand 1–480th part of their volume at 32°. This uniformity in the expansion of gases is attributed to their having no cohesion.

152. EXCEPTIONS.—Water contracts when heated from its freezing point till it reaches the temperature of 39½ (39.38 Halestrom.) Dr Hope found, that in any quantity of this fluid, the coldest water was always at the top when its temperature was below this point. At a higher temperature water obeys the usual law—the hottest is always found at the top.

153. Secondly, Rose's fusible metal contracts when heated from 110.75 to 155.75.

154. Thirdly, Clays and other substances contract when heated, but they do not regain their original volume on cooling.

155. Fourthly, Many fluids expand as they pass to the solid form. Water, iron, bismuth, antimony, and a number of salts, exhibit this in a remarkable manner. Metallic fluids, which have this property, take nice and accurate impressions when cast in a mould, insinuating themselves into all its crevices. The force with which this expansion takes place is very great. By the expansive force of freezing water, the Florentine academicians burst a brass ball; the cavity filled with water was only one inch in diameter, and the force exerted was calculated by Muschenbroek to have been equal to a weight of 27.820 lb.

156. THERMOMETERS.—Our sensations enable us to judge only of a very limited range of temperature, and even there they often deceive us, in consequence of the state of the body at the time; the air of a temperate apartment will feel warm to any one previously exposed to cold, and cold if he has been exposed to heat. The thermometer communicates more precise and definite information. Substances at the same temperature produce no change on it; placed in a warmer or colder medium, the amount of expansion or contraction that ensues indicates the difference of temperature.

157. In making the common thermometer, the ball at one extremity of the glass-tube is heated, and the other end plunged into mercury; this fluid rises as the air in the interior cools, and partially fills it; on boiling it, and plunging the stem as before into mercury, it is entirely filled with this fluid, and the flame of the blowpipe directed against the open extremity, melts the glass, and thus SEALS THE TUBE HERMETICALLY.

158. To GRADUATE the thermometer, place the ball and stem in a mixture of ice and water, and mark the point where the mercury cuts the stem; then place the ball and stem in water boiling in a metallic vessel, the barometer being at 30°, marking the stem as before. Divide the space between these marks into 180°, and extend degrees above and below the original marks, each being of the same size with the intermediate divisions.

159. The ZERO or commencement of Fahrenheit's scale is 32° below the freezing point of water, that having been considered by him as the lowest possible temperature; when he discovered that this was an error, a series of descending degrees were continued below his zero with the prefixture — (minus).

160. In the CENTIGRADE thermometer, the space between the freezing and boiling points is divided into 100°; in REAUMUR'S into 80°.

161. Mercury is preferred in making thermometers from the range of temperature which it commands, and the comparative equality of its expansions from equal increments of heat, while the increased capacity of the thermometer tube almost exactly balances its increasing rate of expansion near its boiling point. Alcohol is employed in thermometers for examining low temperatures.

162. AIR was employed by Sanctorio, who invented the thermometer, instead of

mercury and alcohol; though more delicate than these in its indications, it is affected by the pressure of the atmosphere, independent of temperature. The DIFFERENTIAL THERMOMETER consists of two glass bulbs fixed to the extremities of a glass tube bent in the form of the letter U, this tube being partly filled with a coloured fluid. Heat or cold applied to both balls at the same time, produces no alteration in the level of the fluid, the air expanding or contracting equally in both. If one ball alone be heated, the liquid descends in its stem, being forced down by the expanded air; if it be cooled, the liquid rises as the air contracts. Thus it indicates only a difference of temperature between the balls, and hence the name.

163. In Rutherford's REGISTER THERMOMETER, the highest temperature that occurs within a given time, however long, is known by a small iron wire or index within the stem, which is pushed forward by the mercury as far as it may expand, and left as it contracts; the instrument is placed horizontally, and to arrange it for a new observation it is merely necessary to shake it gently till the index falls down to the mercury. The lowest temperature that may have occurred within any given time is indicated by the descent of a small piece of enamel in the stem of a spirit thermometer, the spirit rising and passing by it as the temperature is increased; the enamel is easily brought to the top of the spirit for a new observation by inverting the thermometer.

164. For measuring high temperatures, Guyton Morveau, and Professor Daniell, each invented a PYROMETER, where the degree of heat is ascertained by the expansion of a bar of platinum.

165. In Daniell's pyrometer, 1 degree is equal to 7 of Fahrenheit. Many valuable results have been obtained with this instrument.

166. Wedgwood endeavoured to measure high temperatures by the contraction which clay suffers when exposed to heat: but a long continued moderate heat produces the same amount of contraction as a higher temperature for a shorter time. It has accordingly fallen into disuse.

167. Metallic thermometers are frequently made for indicating ordinary temperatures, by soldering together metals of unequal expansibility; the compound bar bends as the temperature varies, and a scale being attached, it is graduated by comparing it with a common thermometer.

168. The thermometer, in indicating temperature, communicates no information as to the total quantity of heat in any substance, but merely its disposition to impart or withdraw heat from any substance in contact with it.

SECT. II. *Liquefaction.*

169. It is believed that all solids can be melted by heat, excepting those compounds which are easily decomposed by this agent. Bodies that are fluid at natural temperatures are regarded as melted solids. The melting point varies in different substances. Mercury melts at —39.5; ice at 32; tallow at 96; potassium at 150; silver at 2233 (Daniell); carbon has not hitherto been fused by the highest temperature to which it has been exposed.

170. In a vacuum, a much lower temperature than usual (—5 ?) is necessary to commence the congelation of water, but the temperature rises to 32 whenever it begins, from the latent caloric evolved—See the succeeding paragraphs.

171. Dr Black pointed out the important fact, that after a solid has been heated to its melting point, a large quantity of caloric is consumed in liquefying it, which does not increase its temperature; this he termed LATENT CALORIC; the liquefaction being completed, more heat increases the temperature, as usual. When the fluid becomes solid again, the heat which it had absorbed during liquefaction is restored without any reduction in its temperature.

172. The terms COMBINED or INSENSIBLE CALORIC are frequently used instead of latent caloric, while that which affects the thermometer is said to be FREE or SENSIBLE CALORIC.

173. Different solids require different quantities of heat to render them fluid: ice requires as much as would elevate the temperature of an equal weight of ice-cold water by 140 degrees.

174. Dr Black considered that caloric enters into chemical combination when it causes a change of form without any change of temperature. Dr Irvine considered the absorption of caloric that attends liquefaction (and also evaporation) not to be the cause, but rather the effect of the change of form, in consequence of which, a larger quantity of heat is required to maintain it at a given temperature. This view, however, does not explain so satisfactorily as Dr Black's the change of form; a more extended series of observations on the capacities of solids and liquids is necessary to enable us to form a correct opinion.

SECT. III. *Vaporization and Spontaneous Evaporation.*

175. VAPORIZATION is the rapid conversion of a fluid into a vapour by heat. The agitation accompanying EBULLITION is produced by the escape of the vapour.

176. The vaporific point varies in liquids as the melting point in solids. Some solids pass into vapour when heated, without previously becoming liquid. Bodies easily converted into vapour are said to be VOLATILE; those that are not, are called FIXED. Sulphuric ether boils at 96; alcohol at 176; water at 212; and mercury at 662.

177. Liquids differ much in the amount of expansion attending vaporization: according to Gay Lussac, water expands to 1696 times its volume from its point of greatest density, alcohol to 659, and ether to 443, the barometer being 29.92, and the temperature 212; thus, some liquids, lighter than water, and more volatile, give heavier vapours, expanding less during vaporization.

178. VAPOURS are highly elastic and compressible, expanding when relieved from pressure, and contracting when subjected to it. They resemble gases, but are less permanent, being easily condensed into fluids.

179. The process of DISTILLATION depends upon these circumstances, and consists in the separation of a more volatile from a less volatile substance by vaporization, the product being collected in a receiver. SUBLIMATION is the term employed when the product condenses in the solid form.

180. Gases are regarded as vapours, formed at so low a temperature that it is difficult to condense any of them, except by great cold or pressure.

181. Dr Faraday succeeded in condensing many gases in this manner, producing them in a tube, hermetically sealed, so that each succeeding portion condensed more and more what had already been formed.

Table shewing the Temperature or Pressure required to liquefy the following Gases.

Gases.	Pressure.	Temp.	Gases.	Pressure.	Temp.
Sulphurous acid,	2	45	Sulphureted hydrogen,	15	50
Cyanogen, .	3.6	45	Carbonic acid, .	36	32
Chlorine, .	4	60	Muriatic acid, .	40	50
Ammonia, .	6.50	50	Nitrous oxide, .	50	45

182. Vapours are expanded by heat according to the same law as gases.

183. Dr Black shewed that a large quantity of heat is consumed during the production of vapours, as in the melting of solids. It becomes LATENT, to use his language; it adds nothing to the temperature, it is not sensible to the thermometer.

184. The quantity of heat required to convert each liquid into vapour is peculiar; water requires more than any other that has been examined, viz. as much as would elevate its own temperature by 1000 degrees, supposing it still to retain the fluid form. The heat consumed during vaporization is restored without any reduction of temperature when the vapours are condensed.

185. VAPORIZATION is much influenced by PRESSURE. *The greater the pressure the more elevated the temperature required to produce vapour, the denser is the vapour produced, and the greater its elasticity.* At 212 (barom. 30) steam has just sufficient elasticity to overcome the pressure of the air, and rise against it. If water be heated in a strong metallic boiler, provided with a cover and stop-cocks, the steam can be prevented from escaping while the heat is still applied.

186. With the stop-cocks open, the temperature of the water never rises above 212, the steam carrying away in a latent form the heat communicated; but when they are shut, more steam is forced into the upper part of the boiler, already filled with vapour, rendering it denser, and imparting to it great elastic power: it is now called HIGH PRESSURE STEAM; and it presses with more force than formerly upon the water in the boiler, which, along with the steam, now acquires a higher temperature.

187. Steam is always at the same temperature with the water from which it rises, and accordingly, *high pressure steam, steam produced at a high temperature,* and *steam of great density*, may be considered synonymous terms.

188. The elasticity of steam, or force with which it presses, increases in a greater ratio than the temperature at which it is produced: thus, if it be—

1*	at 212	4	at 293.7	16	at 398.48
it is 2	250.5	8	341.78	24	435.56

189. Or, in other words, steam, at these different temperatures, has 2, 4, 8, 16, and 24 times the elastic power of steam at 212. Hence the frequent explosions of boilers not properly provided with safety valves, and the fatal accidents which ensue.

190. The safety-valve is a metallic plate loaded with weights, and fitted to an aperture in the boiler, so as to prevent the steam escaping, but not so heavily loaded as to prevent it from it from being lifted, and permitting any excess of steam to be discharged, before its elasticity is increased so much that the bursting of the boiler is endangered.

191. High-pressure steam allowed to escape from a boiler instantly expands, and with such rapidity, that it falls much in temperature, and does not scald the hand as steam at 212 does. Great expansion attends the escape of the steam, which is accompanied by a reduction of temperature.—See 259.

192. A REDUCTION OF PRESSURE facilitates the production of vapour. Water boiling in a flask which is corked tightly, while the upper part is full of steam, continues to boil though the heat be withdrawn; the cold heat promotes the ebullition more by condensing the superincumbent steam, and thereby relieving the water from pressure, while the cork excludes the air, than it opposes the reducing the temperature of the water. By plunging the flask in cold water, the same phenomena are seen in a still more striking point of view; but if the flask be put into boiling water, the ebullition immediately ceases, the warm fluid not condensing the steam above the water in the flask.

193. Even the heat of the hand is sufficient to make water boil in a glass tube partly filled with it, when the air has been expelled by boiling, and the open end hermetically sealed; the toy called the WATER HAMMER illustrates this well;

* Equal to the pressure of one atmosphere, according to the common expression.

its name is derived from the noise the water makes in falling rapidly from one end to the other, no air being present to resist its progress.

194. According to Professor Robison, fluids boil at 140 degrees below their usual boiling point, when entirely relieved from pressure. Hot water, with a thermometer in it, and placed under the receiver of an air-pump, illustrates well the nature of ebullition under a reduced pressure; as the air is exhausted the water boils, falling rapidly in its temperature at the same time. No caloric being supplied to it from without, part of the heat it contains is consumed in effecting the change of form.

195. *The quantity of heat that becomes latent during vaporization varies with the temperature at which it takes place; the greater the temperature, the less the latent heat.*

196. According to numerous experiments, it appears that the total quantity of heat in steam as it is formed is always the same; the temperature of vaporization being 112, the latent heat required is 1100; at 212 it is 1000; at 312 it is 900, &c. &c. Hence there is no economy of fuel in evaporating at a low temperature, as was once believed, though the process is highly advantageous, by allowing syrupy solutions to be concentrated, extracts to be made, and many fluids to be distilled at such a moderate temperature as prevents any risk of decomposition.

197. Mr Howard and Mr Barry have made the most important applications of this principle. Sometimes the vapour is withdrawn from the boiler by a powerful air-pump worked by a steam-engine. On other occasions, a large receiver is connected with the still, filled with steam, which is then condensed, and a communication now being opened between the still and the receiver, any air in the still divides itself between the still and the receiver; by repeating this several times, the air in the still can be exhausted to any extent, and a sufficient reduction of pressure continually maintained by cooling the receiver.

198. Vaporization is also influenced by the nature of the vessel in which it takes; water boils at 212 in metallic vessels, but at 213 or 214 in vessels of glass. Powder thrown into heated liquids promote much the discharge of vapour.

199. The principal applications of steam are, 1. As a moving power in the steam-engine; 2. For the communication of heat; 3. As a chemical agent in numerous operations.

200. All vapours obey the same general laws as steam, though the temperature at which they are formed, the quantity of heat that becomes latent, the amount of expansion, &c. may vary in each.

201. Spontaneous evaporation is the term employed to denote the production of vapour by some natural agency, without the direct application of heat. This evaporation takes place to an enormous extent at the surface of the earth and of the ocean, a kind of natural distillation continually going on, by which all the vapour is produced that is returned in the form of rain, snow, hail, dew, hoar-frost, &c.

202. Many fluids evaporate spontaneously as well as water, and even some solids obey the same law, as ice, camphor, and carbonate of ammonia.

203. Evaporation, it is obvious, will cease altogether when the power of gravity between the particles is greater than the action of those forces that facilitate the production of vapour. Dr Faraday's experiments and observations have led to the belief that vapour, even of an exceedingly low tension, is not continually emanating from all substances, as many had imagined.

204. Cause of evaporation. Caloric must be considered as the cause of spontaneous evaporation, as well as of Vaporization. Dalton's experiments have shewn that the actual quantity of vapour which can exist in a given space is dependent upon temperature alone. The presence of air may affect the rapidity of

evaporation, but it neither increases nor diminishes the absolute quantity of water which can be taken up in a given space.

205. **Great cold** is produced during **spontaneous evaporation**, the quantity of caloric that becomes latent being greater and greater the lower the temperature at which it takes place. See 196. Volatile liquids, as ether and liquid sulphurous acid, are often used to produce intense cold with rapidity, as they evaporate.

206. Spontaneous evaporation is much influenced by a number of circumstances, 1. *The volatility of the evaporating substance;* 2. *the temperature;* 3. *the pressure;* 4. *the air;* 5. *the extent of surface presented during evaporation.*

207. **Volatility.** The more volatile the evaporating substance, the more rapidly does it evaporate, at any given temperature or pressure.

208. **Temperature.** The higher the temperature the more rapid is the evaporation, and the greater the quantity of vapour which can be formed in a given space. The **tension, force,** or **elasticity** of vapour, which signifies its power of resisting pressure, depends upon the temperature; it is obvious, therefore, that the higher the temperature the denser and more elastic will the vapour be, in the same manner as at temperature above 212.° See 189.

209. The following examples from a table by Dalton, shew the elastic force of the vapour of water at different temperatures, expressed in inches of mercury; the 3d column shews that each successive increase of temperature from about 20° to 30° doubles the elasticity of the vapour. Dr Ure has published a table on the same subject, and on the elasticity of other vapours. The elasticity of vapour at temperatures below 212°, is most easily shown by putting a little water into a barometer-tube, and then heating the tube to the required point; the depression of the mercury indicating the force of the vapour.

Temp.	Elasticity.	Increase of Temp.	Temp.	Elasticity.	Increase of Temp.
32	0.200	...	145	6.53	26
52	0.401	20	173	13.02	28
73	0.796	21	205	26.13	32
96	1.63	23	212	30.00	...
119	3.25	25			

210. **The elasticity of different kinds of vapour** is not the same, at equal distances above or below their boiling points.

211. **Pressure.** The temperature being constant, the quantity of vapour that can exist in a given space is always the same, according to Dalton, however much the pressure may vary. If the pressure on any vapour resting on the fluid that produces it be diminished, it immediately expands, and evaporation proceeds rapidly till each part of the space above the fluid be charged with vapour as before; if the pressure be increased, part of the vapour is condensed into the liquid form as it is pressed into a smaller space, leaving the superincumbent space as much charged as at first. Accordingly, the more any fluid is relieved from the pressure of superincumbent vapour, the more rapidly does it evaporate.

212. Sir John Leslie shewed that water could be frozen by the cold produced during its own evaporation, when relieved from pressure; for this purpose it is put into a porous cup, over a glass plate containing sulphuric acid, placed in the receiver of an air-pump; the air being exhausted, the acid condenses the watery vapour so rapidly, that more is continually produced, and ice formed in the cup. Dr Wollaston's **cryophorus** or frost bearer (κρυος, cold, and φερω, to bear), illustrates beautifully the same position; it consists of two hollow glass balls connected by a bent glass-tube, one of them containing a portion of water, the apparatus having been sealed hermetically when the air had been expelled by boiling the

water. If the other ball be put into a freezing mixture of salt and snow, the vapour it contains is instantly condensed, and the evaporation from the water in the first ball now proceeds so rapidly, that a pellicle of ice is almost instantly formed on its surface.

213. Ether placed in the receiver of an air-pump, boils rapidly as it is exhausted; it is at ordinary temperatures *much above its boiling point when freed from the pressure of the air;* as the pressure diminishes the ebullition increases, which produces much cold, part of the heat in the ether being consumed in effecting the change of form. Water put in a tube which is placed in the ether is readily frozen; mercury also may be frozen on the same principle.

214. Extent of surface. Vapour is formed only at the surface of fluids, and with greater and greater rapidity the more extensive the surface they present; there is also less impediment from any accumulation of vapour where the surface is extensive, than when confined within a smaller space.

215. Influence of the Air. Evaporation is retarded by the presence of air. If air be loaded with moisture, it is altogether arrested; if the air be dry, it goes on comparatively with rapidity. A rapid current of air promotes evaporation, sweeping away superincumbent moisture. The higher the temperature the more rapid the evaporation in air.

216. In common language, the air is said to contain the moisture associated with it, but the quantity present is dependent solely upon the temperature.

217. Air is more charged with moisture in summer than in winter, and influences numerous operations accordingly; the deep blue colour which the sky presents in summer is attributed to the moisture present.

218. Hygrometers are instruments for determining the condition of the air with respect to moisture. Daniell's dew-point hygrometer, or some modification of it, is most generally used at present; the quantity of moisture in the air is inferred from the reduction of temperature necessary to produce a deposition of dew from the air. The less the reduction required, the greater the quantity present. Mr Jones, and also Dr Coldstream, employed a very simple form of this instrument; it is made of a delicate mercurial thermometer, with a dark glass bulb covered with muslin, excepting a small part of its surface, on which the deposition of dew may be seen as its temperature falls; the necessary reduction of temperature is produced by moistening the muslin with ether. The temperature indicated when the moisture appears is the *dew-point.*

219. Many prefer operating in another manner, pouring water into a glass, and observing with a thermometer how cold it must be before any moisture is deposited on the outside.

220. Sir John Leslie's hygrometer consists of the differential thermometer having one of the balls covered with muslin, so that it can be easily moistened; the less moisture in the air the more rapid is the evaporation from the muslin, and the greater the reduction of temperature.

221. The quantity of moisture in air has also been examined by observing the length to which a human hair, or some similar substance, can be stretched at a given temperature, and also by removing the moisture altogether by chloride of calcium or sulphuric acid. But none of these methods are now much resorted to.

222. Deposition of moisture from the air. A reduction of temperature, or an increase of pressure, causes a deposition of moisture from air previously saturated. The intermixture of strata of air at different temperatures, both previously saturated with moisture, is a very frequent cause of the precipitation of moisture. Dr Hutton proved this experimentally on the small scale, and pointed out its frequent occurrence in nature; it depends upon this circumstance that the

power of air to dissolve moisture increases in a higher ratio than the temperature; and when a mean temperature is established by the intermixture of such strata, as when opposing currents meet, a portion of the moisture is necessarily precipitated. Thus, for illustration, if we suppose

The quantity of moisture in air at	32	to be	200
....	72	...	700
The mean of these is, . . .	52	and	450

But when the air has attained the mean temperature of 52, it cannot retain the mean quantity of moisture 450 parts, part therefore is precipitated.

223. Besides being deposited in the form of RAIN, SNOW, HAIL, DEW, and HOAR-FROST, moisture is also separated from the air in the form of VESICLES, constituting CLOUDS and MIST. The precise nature of these vesicles, the manner in which they are produced, and in which they form rain, &c. is altogether unknown.

224. RAIN occurs most frequently where the currents of air are very irregular. The nearer the equator, the greater the annual fall of rain, but the less the number of rainy days. At Granada, 12° North Lat., the annual fall is 126 inches; in Great Britain, Dr Thomson calculates the average fall at 32 inches.

225. The RAIN GAUGE, by which the fall of rain is estimated, consists of a vessel intended to contain the rain, connected by a tube with a funnel-shaped opening of a determinate size, which thus receives the rain that falls upon a measured surface.

226. When air charged with moisture is heated, and ascends in the atmosphere, it at last becomes so cold that much of the moisture is deposited; in this way hoar-frost accumulates in large quantity on the tops of elevated mountains.

SECT. IV. *Incandescence.*

227. Solids and liquids, when heated to 800° or 810°, appear red in the dark; a temperature of 980° or about 1000° is necessary to produce a similar effect in day-light. This luminousness has been termed INCANDESCENCE or IGNITION, and is quite independent of combustion. All bodies which can present this appearance become luminous at the same temperature.

228. GASES, according to Wedgwood's experiments, cannot be rendered incandescent unless in a state of combustion, though heated so that solids suspended in them immediately become so.

229. Incandescent bodies at first give a dull red light, which gradually brightens till it assumes an orange tint, and ultimately presents a shining white light. The manner in which heat produces incandescence is unknown. See Light.

SECT. V. *Effect of Caloric on Chemical Action.*

230. Caloric promotes chemical action when it diminishes cohesion, as when solids are melted by heat. It opposes chemical action when it increases elasticity, except in the decomposition of compounds containing gases, the elasticity which it communicates to them being often sufficient to expel them from such combinations. Lastly, caloric is also believed to exert a *peculiar effect* in particular cases in promoting chemical action. See Chemical Action.

CHAP. II. COMMUNICATION OF CALORIC.

231. Caloric can be communicated in two very different ways from one kind of matter to another, by Conduction and by Radiation. The greater the difference of temperature, the more rapidly does the communication of heat proceed.

SECT. I. *Conduction of Caloric.*

232. When caloric passes slowly from one particle of matter to another, as in an iron rod, one end of which is heated in the fire, it is said to be CONDUCTED

along the particles of the iron, and the process is termed the CONDUCTION of caloric. Bodies vary much in their power of conducting caloric. In general, the densest substances are the best conductors. Solids are better conductors than liquids, and liquids than gases. A man can remain for a considerable time in air heated to near 300° without much inconvenience, but solids and fluids even at a much lower temperature burn him severely.

233. CONDUCTION IN SOLIDS. Among solids, metals are the best conductors. According to Ingenhousz, silver is the best conductor; then copper, gold, tin, iron, and lastly lead. Despretz has given the following table, but platinum is generally considered to have a much less conducting power than he has assigned to it.

	Conducting Power.		Conducting Power.		Conducting Power.
Gold, . . .	100.	Iron, . . .	37.41	Marble, . . .	2.34
Platinum, .	98.1	Zinc, . . .	36.37	Porcelain, . .	1.22
Silver, . .	97.3	Tin, . . .	30.38	Brick Earth, .	1.13
Copper, . .	89.82	Lead, . . .	17.96		

Stones, glass, and different minerals, come next in order; and lastly, light, spongy, and porous solids, as charcoal, hair, furs, &c.

234. The following table, by Count Rumford, shows the time a thermometer took to cool 135° when surrounded by various substances, that being the worst conductor in which it took longest time to cool; in air, the thermometer cooled through the same range of temperature in 576 seconds.

	No. of Seconds.		No. of Seconds.		No. of Seconds.
Spun silk, . .	917	Sheep's wool, .	1118	Eider down, .	1305
Fine lint, . .	1032	Raw silk, . .	1284	Hare's fur, . .	1315
Cotton wool, .	1046	Beaver's fur, .	1296		

235. Imperfect conductors are in daily use for clothing, to prevent the heat of the body from being too rapidly withdrawn by the cold air; to build furnaces, that the heat may not be withdrawn in a similar manner, bricks being employed for this purpose, and various mixtures of clay with sand and charcoal; and also to preserve ice from the heat that would otherwise be communicated by the external air.

236. The most perfect conductors produce the STRONGEST SENSATION when they are warmer or colder than the body; when warmer, they communicate heat more rapidly than bad conductors; when colder, they take it away more quickly. Iron, marble, and woollen stuffs, at different temperatures, afford good examples of this circumstance.

237. IMPERFECT CONDUCTORS are extremely apt to be broken from sudden alterations of temperature; the heat or cold applied expands or contracts rapidly the portion touched, and the rest remains rigid, no caloric being conducted to it, —a rent is the consequence.

238. MR ARTHUR TREVELYAN has discovered a very interesting phenomenon connected with the communication of heat between metals. In numerous cases where metals at different temperatures are brought in contact, more particularly when a bar of hot brass rests upon a hollow cylinder of lead, the superincumbent metal is thrown into a rapid vibratory motion, which is frequently accompanied by very powerful musical tones, particularly when there is a groove in the lead or in the bar. The cause of the vibrations has not been agreed upon; many have attributed it to the sudden expansion and contraction of the lead at the point of contact. Professor Forbes considers it is produced by a peculiar repulsive action communicated by the heat.

239. CONDUCTION OF HEAT IN LIQUIDS. Liquids and gases are very bad conductors of heat, when it is applied in such a manner that no movement is induced

in their particles by the expansion which takes place at the same time. So feeble is the conducting power of most liquids, that it is easy to boil a portion away from the surface in a long tube, while the rest, with the exception of a small portion at the top, shall not have been at all altered in its temperature.

240. But when heat is applied to the lower portion of any vessel containing a gas or liquid, it is immediately expanded there, and becomes specifically lighter; it rises consequently through the superincumbent portion, while the denser part descends to the source of heat, rising in its turn as it receives caloric and expands. Thus ascending and descending currents are established, and go on continually, till an equilibrium of temperature is established. If cold be applied to the upper portion of any warm liquid or gas, similar phenomena necessarily take place, descending currents commencing the movement, and pushing up the lighter and more expanded fluid.

241. No current ensues, in general, when heat is applied to the upper portion of any fluid, as, by the expansion it produces there, it tends to keep that portion still more than before in its first position.

242. The currents induced in liquids and gases by heat explain many natural phenomena, and have been taken advantage of in numerous operations of art. Thus hot-houses and other apartments are often heated advantageously by hot water rising from the *upper part* of a boiler filled with this fluid, and circulating in pipes, which terminate in a descending tube that carries the water to the *lower part* of the boiler, when it is heated again as at first. The ascent of smoke or heated air from a common fire or furnace depends upon a similar cause; the cold air which supports the combustion forming the heavy descending current which pushes up the lighter and warmer air in the chimney.

243. When air, water, sand, mercury, or various solutions, are heated so that bodies placed in them are subjected to an equal and regulated temperature, they are said to be heated by an AIR, WATER, SAND, or MERCURIAL BATH.

SECT. II. *Radiation of Caloric.*

244. When we stand in the sunshine, or opposite a fire, we feel the heat which they communicate. But it does not travel slowly, as when it is conducted: it passes with the greatest rapidity through the air. In this case it is termed RADIANT CALORIC, and is believed to move with the same velocity as light, viz. 192,000 miles per second. Caloric does not affect the air through which it radiates, nor is it affected during radiation by any current of air; it does not radiate through opaque substances, and it is still disputed how far it radiates through transparent solids or liquids. It appears to pass through the most perfect vacuum in the same manner as through air; the heated body also, not being cooled in a vacuum by conduction through the air, may appear from this cause to radiate more than usual.

245. Caloric is emitted in every direction from hot bodies, in the same manner as the rays of the sun. The effect it produces diminishes as the square of the distance; if the effect be reckoned 1 at the distance of 1 foot, it will be 4 times less at 2 feet, 9 times less at 3 feet, 16 times less at 4 feet, 25 times less at 5 feet, and so on.

246. Radiant caloric falling upon solids or liquids must be REFLECTED, ABSORBED, or TRANSMITTED. Sir John Leslie and Count Rumford proved that the nature of the surface of bodies has a most important influence upon the radiation of heat, and also upon the effect produced by radiant heat. ROUGH and POROUS SURFACES radiate powerfully when warm, and absorb with equal power any caloric that falls upon them. BRILLIANT and POLISHED METALLIC SURFACES reflect heat powerfully, but radiate very feebly. Professor Forbes has shewn that heat

may be POLARIZED, or acquire peculiar properties in the same manner as light by reflection and refraction.—See Light.

247. The radiating power of a surface coated with lamp-black being 100, the radiating power of paper is 98, glass 90, and of a polished metallic surface only 12.

248. Caloric, like light, falling on a polished metallic surface, is reflected at an angle equal to the angle of incidence; accordingly, a parabolic reflector gives off parallel rays when a heated object radiating towards it is placed in its focus; and if a similar mirror be opposed to the first, it will collect these rays in its focus, producing a powerful heat. Many of the most interesting facts connected with this subject were ascertained by experiments made with reflectors: the power of different surfaces in radiating caloric was determined by opposing to a reflector with a thermometer in its focus, the different sides of a square tin cannister containing boiling water, after coating them with the substances required.

249. Radiant caloric discharged from any substance at a temperature comparatively low, is intercepted in a great measure by a pane of glass: when the temperature is higher, a great portion of rays are transmitted through the glass, according to De la Roche; but Leslie and Brewster maintained that the effect supposed to depend upon the radiation of caloric through the glass, arises from its being absorbed by one side, conducted towards the other, and then radiated from it.

250. The calorific rays of the sun penetrate glass with facility. They are considered to be capable of passing through glass without heating it.

251. Radiant caloric proceeding from luminous bodies produces a greater effect on different substances the darker the colour.

252. When any cold substance is placed in the focus of a polished metallic reflector, and a thermometer in the focus of another opposed to it, the temperature immediately falls: this apparent radiation of cold is produced by the thermometer radiating caloric towards the cold object.

253. In a clear night, all good radiating surfaces fall much in their temperature, and dew is deposited upon them, as Dr Wells pointed out, in consequence of their being so cold, and condensing the moisture in the air. Occasionally the temperature of the air next the ground is many degrees below the temperature of the air a little higher, from the same cause. In a cloudy night, there is little or no reduction of temperature from radiation.

254. Sir John Leslie's AETHRISCOPE shows beautifully the power of the clouds in preventing radiation. It consists of the differential thermometer, having one of the balls excluded from the light, and the other placed in a polished metallic cup. Exposed to a clear part of the sky, the heat radiates from it rapidly, and the temperature falls; exposed to a cloud, the radiation is returned, and there is no reduction of temperature.

255. Radiation is attributed to the repulsion exerted by the particles of caloric, and the tendency of this power to distribute itself till an equilibrium of temperature is established. PICTET was of opinion that caloric is radiated solely from the warmer to the colder substance; PREVOST maintained that all substances radiate to each other, according to their temperature, and that an interchange still takes place even where an equilibrium has been established; this opinion has been more generally received.

CHAP. III. SPECIFIC CALORIC.

256. All bodies do not contain the same quantity of caloric at the same temperature, nor can they be elevated in their temperature to the same extent by communicating to them the same amount of caloric. For example, if equal quantities of caloric be added to water, oil and mercury, for every degree that the water

rises, the oil will rise 2 and the mercury 23. The term SPECIFIC CALORIC was introduced to express the comparative quantity of caloric peculiar to each substance, which is required to produce a given elevation of temperature. The term CAPACITY FOR CALORIC is frequently employed with the same view, but not so much as formerly. Boerhaave and Black made the first experiments on this subject.

257. Dulong and Petit's experiments have led to the conclusion that the atoms of elementary bodies have the same specific caloric, or require the same amount of heat to be elevated in their temperature by the same number of degrees.

258. Bodies are considered to have a greater capacity for caloric in the liquid than in the solid form; but additional experiments are required on this subject. The relative capacities of vapours and fluids have not been determined; the researches on this subject by Crawford, by Berard and De la Roche, and other experimenters, do not agree.

259. The capacity of bodies for caloric increases in a greater ratio than the temperature.

260. When gases or vapours are relieved from pressure, they expand, their capacity for heat increases, their temperature accordingly falls. Hence the reduction of temperature observed at the tops of mountains, in the air of the air-pump, as it is exhausted, and in steam issuing from a boiler where it has been much compressed.

261. When air is compressed, its capacity is diminished, and heat is evolved, which renders it warm. Thus a change of capacity, without any change in the quantity of caloric, is accompanied by a change of temperature.

262. By CHEMICAL ACTION the capacities of bodies for caloric are generally altered, and as the evolution or absorption of heat is not always proportional to the change of capacity, this circumstance affords an argument in favour of Dr Black's views of combined caloric.

263. Dr Irvine, believing the total quantity of heat in bodies to be proportional to their capacity, endeavoured, on this principle, to ascertain the ABSOLUTE ZERO, or that reduction of temperature at which bodies are entirely destitute of caloric. The great discordance in the results obtained by different experimenters, have shown that this method cannot be depended on.

CHAP. IV. DISTRIBUTION OF CALORIC.

264. Bodies not only differ in the quantity of heat they contain at the same temperature, as has been shown in the preceding chapter, but also to a very great extent in their actual temperature, according to the position they occupy in the globe itself. The heat accumulated at the equator by the sun's rays, produces an ascending current of air to either pole, while other currents of air with less heat pass along the surface of the earth from the poles to the equator. Similar currents take place in the ocean. They all acquire an oblique direction from the rotation of the earth; their course is also modified by numerous local circumstances.

265. The currents of air that pass towards the poles are not at a high temperature, as the expansion attending their ascent in the atmosphere is necessarily accompanied by an increase of capacity. This does not prevent them, however, assisting in equalizing the temperature of the globe, as their capacity must be diminished, and their temperature increased, when they descend in the polar regions.

266. The rays of the sun produce little or no direct effect upon the air through which they pass, but are absorbed at the surface of the earth, and then communicate heat by conduction to the surrounding air. Penetrating less deeply on the surface of the land than in the water, it is apt to acquire a higher temperature

than the water by day, but becomes proportionally cold by night, from the loss of heat by radiation. The air on the land being warmed accordingly to a greater degree during the day than the air on the water, a SEA BREEZE proceeds towards the land, the heavier pressing in upon the lighter air; during the night the effect is reversed, and a LAND BREEZE prevails.

267. In INSULAR SITUATIONS, extreme heat and cold, such as prevails in large continents, are moderated to a very great degree by the water of the ocean, which is much more equal in its temperature than the air. All currents of hot and cold air are rendered milder by passing over an extensive surface of water.

268. The MEAN TEMPERATURE *of the air at the surface*, in different parts of the globe, varies probably nearly to the extent of 100° F. At the equator it is 82.5; at Rome, latitude 41.53, it is 60°; at Edinburgh, latitude 55.57, it is 47°; at Melville Island, latitude 74.45, it is —1.5; and at the Pole, it has been calculated by Arago to be —13.

269. At a certain elevation in the air, varying in different latitudes from the Equator to the Arctic Seas, the temperature is always 32°. This has been termed the LINE OF PERPETUAL CONGELATION; at the Equator it is 14.760 feet; at the Alps it is 8220; in Iceland 3084; and at the Polar regions perpetual frost reigns at the surface of the earth.

270. The MEAN TEMPERATURE OF THE SURFACE OF THE OCEAN varies from 32½° in the North Sea to 82° at the Equator. But at great depths the mean temperature of the ocean varies only from 31° to 44°. Between the North Sea and the Equator, and at the bottom of large lakes, nearly the same variation has been observed.

271. The varying temperature at the surface affects the soil only to the depth of a few feet. From numerous geological phenomena, and experiments made in mines, it is believed that the temperature at great depths is very elevated.

CHAP. V. PRODUCTION OF HIGH AND LOW TEMPERATURES.

SECT. I. *High Temperatures.*

272. The different sources of heat are, the Rays of the Sun, Mechanical Action, Electricity, and Chemical Action.

273. RAYS OF THE SUN.—When they are concentrated by a lens, or parabolic reflectors made of polished metal, a most intense heat is produced, which can be applied in a manner often very advantageous in numerous cases of chemical action.

274. MECHANICAL ACTION.—An elevated temperature may be easily excited by compressing air in a syringe, rubbing pieces of dry wood, or by striking them with a hammer. Condensation always takes place, producing a diminished capacity for heat; unless the condensation be increased on each successive trial, or the material operated on be previously brought back to its original condition by the application of heat, so that it may be condensed again as at first, there is no evolution of heat.

275. ELECTRICITY is a powerful agent in producing a great elevation of temperature. See Electricity.

276. CHEMICAL ACTION is the great source of heat for artificial purposes, as in the combustion of coal and fermentation. During chemical action, an evolution of heat may proceed from a change of capacity, or from the separation of heat previously in a state of chemical combination. The various methods of producing heat by chemical action will be described under the history of individual substances.

Sect. II. *Low Temperatures.*

277. Low temperatures may be produced in many different ways.

278. I. By PROMOTING RADIATION, as when a substance with a rough porous surface is exposed to the air in a clear cloudless sky.

279. II. By DIMINISHING THE PRESSURE ON GASES, as when they expand, and their temperature falls as their capacity increases.

280. III. By PROMOTING EVAPORATION, either by diminishing the pressure on a liquid while evaporating, or passing a rapid current of air over its surface, so as to remove quickly any superincumbent vapour.

281. IV. By CHEMICAL ACTION, and more especially by the rapid liquefaction of solids. When solids are rendered fluid by chemical action, without the application of heat, part of their caloric becomes latent, and hence the reduction of temperature. See Table of Freezing Mixtures.

III. LIGHT.

282. Light is regarded by many as a distinct and peculiar matter, emitted in rays from all luminous and visible objects which are not black, in the same manner as heat. Others consider it more probable that it is produced by vibrations or undulations in an extremely attenuated and elastic ether, diffused throughout the universe. The phenomena which it produces are usually explained in the language of the first opinion; but the latter is now gaining ground, and is perhaps that which is most generally received at the present moment.

283. Light moves with immense velocity, viz. 192,000 miles each second. It passes through perfectly TRANSPARENT COLOURLESS bodies with little or no diminution: those that are WHITE reflect it; those that are BLACK absorb it; and those that present various coloured tints retain part and give out the rest.

284. Light is bent towards the perpendicular, when it passes from a rarer to a denser medium at an oblique angle, as on entering water after passing through the air: this is termed REFRACTION.

285. When a ray of light passing through any object is divided into two parts or pencils, each of which produces a distinct image, it is said to suffer DOUBLE REFRACTION.

286. When a ray of light acquires peculiar properties by refraction or reflection, so that it enters transparent bodies at certain angles and not at any others, it is said to be POLARIZED, and the change is termed POLARIZATION.

287. WHITE LIGHT, according to the analysis of Newton, consists of seven different kinds of rays, viz. *red, orange, yellow, green, blue, indigo, and violet.* The decomposition of white light is effected by passing it through a glass prism; the component rays being refracted unequally, are separated from one another in the order mentioned, the red or least refrangible ray following most closely in the direction the ray would have pursued had it not been refracted at all, and the violet ray deviating most from its direction. The coloured band produced in this manner is termed the PRISMATIC SPECTRUM.

288. According to a new analysis of light by Sir David Brewster, he considers that, red, yellow, and blue light, exist in every part of the spectrum, and that all the other tints are formed by these primary colours, the red predominating where the least refrangible rays occur, the yellow in the middle, and the blue where the most refracted rays are seen.

289. When the rays of light as they are separated by a prism are collected again by a lens, white light is produced.

290. The greatest illuminating power is found in the brighest yellow or palest green.

291. The SOURCES of light are the Sun's Rays, Chemical Action, Electricity, Phosphorescence and Crystallization.

292. In the SUNBEAM, three different kinds of rays have been detected, viz. CALORIFIC, LUMINOUS, and CHEMICAL RAYS. The calorific rays were first noticed, altogether apart from those of light, by Sir W. Herschel: he found them beyond the red or least refrangible ray. Subsequent experiments proved that their position was considerably influenced by the kind of prism employed, but that they are least powerful in the violet ray, and increase in general towards the red ray or beyond it.

293. The chemical rays are met with beyond the violet ray, towards the other extremity of the spectrum, and are therefore the most refrangible rays in the sun beam: they were noticed by Wollaston and Ritter. The calorific and chemical effects of the sunbeam are now attributed more to the action of those separate rays which accompany the luminous rays, than to any direct effect of the luminous rays themselves. Morichini, and also Mrs Somerville, have stated that the violet rays can communicate magnetic properties to iron: this subject is still under examination.

294. Dark coloured bodies are more powerfully affected by the heat of the sun than others, though radiant caloric free from light, in the opinion of many philosophers, does not affect them more than other substances: Dr Stark, however, has inferred from his late experiments, that the absorption of heat both from non-luminous and luminous bodies, and also its radiation, is much influenced by colour. In looking at the sun through glasses of various colours, the eye suffers much from heat at times, while the light is feeble, according to the coloured glass employed.

295. The CHEMICAL EFFECT of light, or of the rays which accompany those of light, is conspicuous in its action upon chlorine and hydrogen, a mixture of which is instantly detonated by a bright sunshine; chloride of silver, in a few seconds, passes from a bright white to a dark colour, when exposed in the same manner. Its power of promoting chemical action is very trifling compared with that which heat exerts, but is too important to be overlooked, especially in the chemical actions that take place in the animal and vegetable kingdom.

296. The evolution of light by CHEMICAL ACTION is too familiar to require any particular illustration; it is well seen in the phenomena of combustion. These, and the INCANDESCENCE of solids and fluids heated to a certain temperature, show how intimate a connection subsists between heat and light, though in what that consists has not been pointed out. Many have adopted the opinion that they are mutually convertible.

297. PHOSPHORESCENCE signifies generally the evolution of light from different bodies called PHOSPHORI, without any chemical action going on, and without that elevation of temperature necessary for incandescence. Many mineral substances, native and artificial, acquire this property when heated and exposed to light, as sulphate of baryta, and also oyster-shells after they have been heated with sulphur. Heat increases the light emitted, but renders it of shorter duration. Mr Pearsall has shown that some bodies not naturally phosphorescent, or which have lost this property by heat, acquire it again when subjected to the electric spark. The glow-worm, and many marine animals, are very phosphorescent. The cause of phosphorescence is not very apparent; it is independent of chemical action, and takes place even in vacuo. Several substances, when a change of form is induced

in them, evolve light, more especially some salts and other bodies at the moment of crystallization.

298. The ILLUMINATING POWER of different lights is ascertained by comparing the depth of shadow which they produce, when some object is interposed between them and a screen, the deeper the shadow the more powerful the light.

299. LESLIE'S PHOTOMETER is constructed on the principle that the intensity of light is proportional to its heating effect when absorbed. It consists of his differential thermometer with one of the balls blackened. A thin glass cover protects the balls from being heated by currents of air. The sun's rays do not affect the clear ball, but are condensed by the blackened ball, and produce a corresponding elevation of temperature. But when this photometer is used with light produced by combustion, or by ordinary incandescent bodies, its indications are inaccurate; the calorific effect of red-hot coke, for example, produces more effect than that of many kinds of light, which are evidently much more powerfully luminous.

300. Dr Stark has lately made some very important observations on the influence of colour on the absorption and disengagement of odorous matters. He found, while engaged in Practical Anatomy, that black clothes absorbed much offensive odorous matter, and retained it powerfully, while a light-coloured dress was affected by it in a very trifling manner, and lost what it did absorb very quickly by exposure to the air. Extending his observations to odorous matter in general, he has found that white-coloured bodies are the least absorbent, and dark-coloured the most absorbent, of odorous matter, and has made several important applications of the result of his experiments, in respect to clothing, white-washing, the retention of noxious effluvia by different coloured bodies, and the consequent communication of disease. London Phil. Trans. 1833.

IV. ELECTRICITY.

301. All bodies contain a peculiar power or matter, which has received the name of Electricity, from ηλεκτρον, the Greek for *amber*, the substance in which some of the phenomena it produces were first noticed. Gilbert, in the 17th century, having shown that numerous other substances could produce similar appearances, drew more attention to the subject, and laid the foundation of Electricity as a science.

302. Electricity is an agent of great force, and affects all kinds of matter. It communicates powers of attraction and repulsion, which are very conspicuous in the mechanical effects it produces. It influences the chemical relations of all bodies, and has been considered by some of the most distinguished chemists as the real cause of all chemical phenomena. It is the source of perhaps the most intense heat and light that can be produced by artificial means. It is the cause of magnetism, and it exercises also the most powerful influence over the animal economy. The attraction of a small piece of paper by amber, after rubbing it with a woollen cloth, and the lightning in the thunder storm, are equally produced by electrical action, as was proved by the celebrated Franklin.

303. It is usually considered as a fluid whose particles repel each other, but attract those of other bodies. It moves with great velocity; attempts have lately been made to measure the rapidity of its progress by Mr Wheatstone.

304. Electricity, like heat, tends to diffuse itself over every kind of matter till an equilibrium is established. It cannot be collected and kept by itself like hydrogen or any ordinary kind of matter, and it is still doubtful how far it ought to be regarded as a mere property of matter, or as a distinct principle; the latter opinion has been preferred. Its properties have been ascertained only by observing the phenomena it produces, as it passes from one kind of matter to another.

305. Electricity, according to the opinion of Franklin, may be regarded as a simple fluid, electrical phenomena being produced by its being accumulated or rendered deficient in different bodies; when in excess, the phenomena are produced by the electricity communicated to surrounding objects; when deficient, they arise as the electricity flows in from those which contain a larger quantity to restore the equilibrium.

306. When a substance has acquired more electricity than its natural share, it is said to be POSITIVELY ELECTRIFIED. When it has less, it is said to be NEGATIVELY ELECTRIFIED; the terms POSITIVE and NEGATIVE ELECTRICITY, though not so correct, are occasionally employed in the same sense.

307. Dufay had previously considered that different bodies contained different kinds of electricity, which he termed VITREOUS and RESINOUS; the former being most easily produced from glass, and the latter from resinous matters. Symmer extending his views, believed that ALL BODIES CONTAIN TWO KINDS OF ELECTRICITY,—that these are separated more or less from each other during electrical action,—that they neutralize each other as they combine, when all electrical phenomena disappear. This view of the compound nature of electricity is still maintained by many, particularly on the Continent.

308. The term *vitreous*, according to this opinion, corresponds with the *positive* electricity of Franklin, and the term *resinous* with *negative* electricity. Franklin's opinion is preferred in this country, as it is equally consistent with the phenomena presented, admits of their being explained in more simple language, and involves less hypothesis than Symmer's extension of Dufay's opinion.

CHAP. I. COMMUNICATION OF ELECTRICITY.

309. The power of different bodies in carrying or conducting elasticity varies exceedingly. Metals are the best conductors, then follow successively charcoal, acid liquids, solutions of salts, animal substances, water, and gases rarefied much by heat or diminution of pressure, or charged with a large quantity of moisture. These bodies, having so much more power than most other substances in communicating electricity, are generally termed CONDUCTORS.

310. The term NON-CONDUCTORS is applied to those bodies which have comparatively a very feeble conducting power, though they are not perhaps to be regarded as entirely destitute of it. The principal NON-CONDUCTORS are sulphur, resinous substances, and many earthy compounds, as glass; also air, wood, paper, hair, silk, and feathers, when they are dry.

311. Dr Faraday has made the very important discovery, that numerous substances which do not conduct when solid, conduct readily when fused by heat.

312. When electricity is to be communicated from one kind of matter to another, a connection is made between them by a copper, silver, or other metallic wire. Plates of metal, tin-foil, and mercury, are of great use in adjusting different kinds of apparatus. By amalgamating metals with mercury, a very intimate degree of contact is insured, so necessary in numerous experiments.

313. The passage which the electricity takes is termed the ELECTRIC CIRCUIT; it is said to move in a current, and always proceeds from the body in which it is accumulated to that in which it is comparatively deficient.

314. When the line of communication is not maintained continuously by good conductors, it is said to be BROKEN or INTERRUPTED, and it is there that bodies are placed which are to be subjected to the influence of this agent, which passes through them as it proceeds from one part of the metallic communication to the other, if they be not too far distant.

315. By passing electricity through air or water during part of the circuit, in-

stead of metal, it is retarded in its progress, and can thus produce many effects which are not otherwise easily induced. This is done by making the metallic communication imperfect, as in the preceding case, so that the electricity in its progress through any metallic chain or wire, shall have to pass through a portion of air or water where the metallic contact is broken.

316. Electricity spreads over the surface of bodies, when it does not produce any chemical effect; it does not penetrate into the interior: in perfectly spherical bodies it is equally distributed over the surface; in others, it accumulates at particular parts.

317. Pointed bodies carry off electricity with much greater rapidity, and from much greater distances, than those of a globular form; they are also equally effectual in discharging electricity when it is accumulated in them. They are continually used for these purposes in electrical experiments. When the object is not so much to carry off a continuous stream of electricity, as to discharge a large quantity at once, then metallic balls are preferred to points.

318. An ELECTRICAL DISCHARGER is commonly made of a brass rod, with a brass ball at either end, supported in the middle by a glass handle; the glass being a non-conductor, it may be held freely in the hand, while the electricity passes by the rod, on connecting it with bodies in different electrical conditions to discharge any excess. Various forms are used.

319. When an electrified body is supported on glass, or any other very bad conductor, it is said to be INSULATED.

320. When electricity in great intensity is produced, as by the electric machine, it is much more easily conducted than that from the galvanic battery, where the tension is very feeble.

321. The difference in the conducting power of various bodies for electricity was discovered by Grey: it is attributed to the different forces with which they attract electricity.*

CHAP. II. EXCITATION OF ELECTRICITY.

322. A body is electrically excited when a portion of its electricity is removed, or an additional quantity communicated to it from some external source; its excitation depends upon its tendency to communicate electricity to, or withdraw it from, surrounding objects. In general this effect is produced by Friction, Induction, the application of Heat, or Chemical Action. Magnetism also can induce electricity, while electricity can in its turn produce magnetism, so that both these powers are referred to the same cause, and regarded as modifications of each other. Whenever the electric equilibrium is restored, all excitation ceases.

SECT. I. *Excitation of Electricity by Induction.*

323. Any body electrically excited, when opposed to another in its natural state, causes it immediately to assume an electric condition, exactly opposed to that in which it is itself. Thus, if one end of a brass cylinder, insulated on glass, be opposed to a body positively excited, that end will become negative, a portion of its electricity being expelled to the other end, which becomes thereby positively electrified.

324. If electricity be introduced into the interior of a jar, its presence will cause a portion in the interior to move away; and if the exterior be surrounded with very bad conductors, so that little or none can be discharged, electricity cannot then be communicated to the interior.

* For some farther remarks on Electricity, see Faraday's Researches, in the section on the Effect of Electricity on Chemical Action.

325. Changes of this kind, effected merely by proximity to an excited body, are called cases of the INDUCTION of ELECTRICITY.

SECT. II. *Excitation of Electricity by Mechanical Action.*

326. Friction is a very common source of electricity. By rubbing glass, sealing-wax, or amber, the rubber assumes one electrical condition, and the glass, or other substance, passes to a different state.

327. The nature of the surface, and of the substance used as a rubber, influences the result. Thus, smooth glass is rendered positive by a woollen cloth, but rough glass becomes negative with the same substance, and sealing-wax is rendered positive by metallic substances, but negative by hare's skin.

328. Bad conductors have been called ELECTRICS, from the facility with which electrical phenomena are induced in them; while good conductors have been called NON-ELECTRICS, because, unless insulated, they are with difficulty rendered electrical.

329. In the common electrical machine, the electricity is produced by the friction of a hair-cushion and a piece of silk with a cylinder or plate of glass, which is made to revolve rapidly; pointed brass wires (see 317) remove the electricity to a large cylinder or ball, made of sheet-brass, in which it is accumulated. This is termed the PRIME CONDUCTOR; from it all the electricity developed by the common electrical machine is procured. It is insulated by a glass support. Were it not insulated, the electricity communicated to it would pass at once to the earth.

330. The rubber must be connected with the ground by a chain or other good conductor, to supply electricity to the machine as it is exhausted, otherwise it will soon cease to be produced; the rubber and the chain are generally connected in the cylinder machine by a conductor, which has been termed the RUBBER'S CONDUCTOR, similar in its form to the prime conductor.

331. The LEYDEN JAR, named from Leyden, where it was invented, is a contrivance by which electricity can be accumulated in larger quantity than in the prime conductor. It is a glass jar or bottle, coated to within a few inches of the top with tinfoil, both on the interior and exterior surface, a rod with a ball at the top passing through the cork to the interior coating. Electricity communicated from the prime conductor to the ball, is transferred by the rod and metallic coating to the interior of the jar; a corresponding quantity leaves the exterior in consequence of induction (324), and when one ball of the discharger (318) touches the exterior of the jar, and the other is approached to the ball of the brass rod, the electricity passes as a powerful spark from the interior to the exterior, and the equilibrium is restored.

332. The LEYDEN BATTERY consists of many jars, connected externally by placing them on tinfoil, or some other good conductor, and internally by a metallic connection attached to the internal coating of each jar; the whole of the internal surfaces thus act together, and also the whole of the external surfaces. By charging one jar the whole are charged, and by discharging it, the whole are discharged at the same moment.

SECT. III. *Excitation of Electricity by Chemical Action.*

333. The excitation of electricity by chemical action has been proved in the most satisfactory manner by Wollaston, Pouillet, Davy, De la Rive, and still more especially by Dr Faraday, who, in a series of Memoirs of Electricity, has elucidated, with the greatest sagacity and ingenuity, many of the most intricate

subjects connected with chemical action, and opened, at the same time, many new and promising fields of investigation.

334. Electricity developed by chemical action is usually termed GALVANISM or VOLTAIC ELECTRICITY, from Galvani and from Volta. Galvani pointed out the first fact connected with this branch of electricity, viz. that contractions are excited in the limbs of a frog when a zinc and silver plate are placed, the one upon the nerve and the other upon the muscles, and then made to touch each other. Volta, not agreeing with Galvani in the opinion that the electricity was present in the muscles, and merely conducted by the metals, supposed that the effect was due to the contact of the metals which produced the electricity, and soon afterwards constructed the VOLTAIC PILE, which is similar in its general arrangement to the trough now in use.

335. But the liquids or moistened solids, used to connect each pair of metallic plates, are not now considered conductors merely, as Volta believed; they are the real source of electricity during their action upon the metals. Davy, it is necessary to mention, considered with Volta, that the first action commences by the contact of the metals, though he admits that it is sustained by the chemical action of the substances employed.

336. The galvanic trough, invented by Cruickshanks, is the apparatus now employed for producing electricity by chemical action. Plates of zinc and copper, 4 or 6 inches square, are soldered together, and cemented in a wooden trough, so that a series of cells are formed, in which the liquid is placed which is to act chemically, and produce the electricity. The plates are arranged so that all the zinc surfaces shall be placed in the same direction. Copper wires are fixed to the extreme plates at each end of the battery, so that they can be easily connected by a good conductor.

337. The liquid employed usually consists of 1 part of acid to 15 or 25 of water, when a very strong charge is required; and of 1 part of acid to 80 or 100 of water for a weak charge. Nitric acid is preferred for the evolution of heat and light, but it is seldom used alone, being mixed commonly with from 1-10th to a-half of sulphuric acid. Saline solutions, and even common water, are sufficient for some purposes, but they are exceedingly feeble compared with the preceding mixtures.

338. According to the above arrangements, the electricity developed by the action of the acid liquid in the first cell and the zinc, the metal most affected, circulates through the liquid to the copper, which communicates it to the zinc in contact with it; as this takes place throughout the whole trough, each succeeding pair of plates handing over, as it were, to the next pair, what they may have produced along with what they have received from the preceding pairs, the electricity becomes accumulated at the last pair, terminating in a copper plate, which has accordingly received the name of the POSITIVE POLE or extremity of the battery, the other being termed the NEGATIVE POLE.

339. SIMPLE GALVANIC CIRCLES consist only of a single pair of plates; in COMPOUND GALVANIC CIRCLES, as in the galvanic trough, there are a number of pairs of plates.

340. The term GALVANIC BATTERY is usually applied where a number of troughs are connected together by copper wires; the same arrangement of the plates being carefully attended to, so that they may act as a single trough.

341. In compound galvanic circles, where soldered plates are used, there is a superfluous zinc plate at one end, and a superfluous copper plate at the other end of each trough; *and hence the zinc plate, in common language, is often termed the positive pole of the battery, though it is so solely in consequence of being next the copper,*

which communicates to it electricity in the same manner that it would give it to the connecting wire, or any other good conductor. Similar remarks apply to the superfluous copper plate at the other extremity.

342. A great variety of arrangements, on the same principle as the trough, have been from time to time employed. The pile originally employed by Volta consisted of the plates of metal arranged in pairs above each other, with a moistened cloth interposed, the moistened cloth representing the saline solution in the cells of the trough.

343. A porcelain trough with cells is often used, the plates being suspended individually in a frame, and a copper and zinc plate introduced into each cell; the zinc plate in one cell is always joined by a metallic connection to the copper plate in the next cell: arrangements of this kind have been known by the name of COURONNE DE TASSES, having been made originally with a number of cups.

344. Sometimes the copper plate is made twice the usual size, and doubled in such a manner that each surface of the zinc has a surface of copper opposed to it; in this way a greater degree of power is obtained, without a corresponding increase in the quantity of zinc employed. When a single pair of plates is used, the copper plate is frequently made so as to contain the liquid, and the zinc, when immersed, is prevented from coming directly in contact with the copper by pieces of glass or baked wood; troughs have also been constructed on the same principle by Mr Hart, which have been found very convenient.

345. Mr Pepys and Dr Hare have employed the metals in coils for some purposes; the latter has termed a battery made of many different coils a CALORIMOTOR, from its great power in deflagrating metals.

346. For a number of experiments, no arrangement is so well adapted as that introduced by Mr Kemp, where, instead of a plate of zinc, a solution of zinc in mercury is opposed to the plate of copper.

347. Numerous galvanic arrangements may easily be made besides those with zinc, acid, and copper, but these, from their convenience, are preferred. The zinc produces the electricity by its action with the acid, and by the copper it is attracted and removed to the pole.

348. De la Rive has shewn that the metal or substance most acted upon becomes positive with respect to the other.

349. The electrical action ceases when the chemical action terminates between the substances employed.

SECT. IV. *Excitation of Electricity by Heat, &c.*

350. When heat is communicated to different substances, their electric equilibrium is disturbed; this has been observed more particularly in various arrangements with the metals. The facts connected with this subject are usually arranged under the title of THERMO-ELECTRICITY, a department of science more particularly connected with Natural Philosophy than Chemistry. Electricity can also be induced by MAGNETISM; some remarks on this subject will be made in the next chapter. Lastly, electricity is not only developed during perhaps almost all chemical operations that take place both in the organic and inorganic world, but there are animals in which a peculiar galvanic apparatus is found, by which it can be accumulated and discharged at will, as in the Torpedo, and Electrical Eel.

351. Melloni's thermo-multiplier consists of many pairs of small slips of antimony and bismuth, soldered together and arranged so as to have their galvanic action induced by heat; the effect is indicated and measured by a magnetic needle placed below, a wire connected with the metallic slips being coiled around it. Variations of temperature, not appreciable by the common thermometer, are

easily observed with this instrument. In experimenting with it, he has found that radiant heat passes directly in various quantities through some bodies. These he has termed Diathermal Bodies, and he has also been led to entertain the opinion that there exist different species of rays of heat as well as of light. This instrument has been much used in Professor Forbes's important experiments on the Polarization of Heat.

CHAP. III. EFFECTS OF ELECTRICITY.

352. The principal effects of electricity are the production of mechanical and chemical action, of heat, light, and magnetism, and lastly, its action on the animal economy.

353. The action of the electric power is much influenced by the intensity, and also by the quantity brought into play. From the electrical machine, electricity of great intensity is procured, but in quantity far inferior to that which is produced by the galvanic battery. From the latter again, though the quantity that may be procured is comparatively great, its tension or intensity is always exceedingly low; the greater the number of plates, the greater the tension; the greater the extent of surface, and the more powerful the chemical action, the greater the quantity of electricity produced in a galvanic battery. For producing heat and light, plates of a very large size are preferred, though few in number; but for chemical decomposition a number of plates are necessary.

Sect. I. *Mechanical Effects of Electricity.*

354. Bodies similarly electrified, positively or negatively, repel each other. Bodies in different electrical conditions attract each other. Electricity produced by friction, as with the electrical machine, is most powerful in communicating these properties to small masses of matter, as pith balls, pieces of paper, and fragments of thin metallic leaves. The galvanic battery is exceedingly feeble in this respect.

355. Electrometers, or measures of electricity, for ascertaining the intensity with which attractions and repulsions of this kind are exerted, have been contrived, the intensity being inferred from the distance to which some substance of a fixed size and form is repelled, when subjected to the influence under examination.

356. The nature of the electrical excitement communicated can also be ascertained by means of an electrometer. Thus, if two gold leaves suspended together be made to diverge to a certain extent, by rendering them positively electrical, and if any substance electrically excited approach them when they repel each other, the divergence will be increased if it be positively excited, and diminished if it be negatively electrified. In Bennet's gold-leaf electrometer, two pieces of gold-leaf are suspended in a glass case from a metallic plate; slips of tinfoil are placed on either side to remove the electricity when the gold leaves diverge so far as to come in contact with them.

357. Electricity, retarded in its course by bad conductors, operates, if it be in sufficient quantity, exactly in the same manner as a powerful mechanical agent, as when lightning is interrupted in its progress towards the earth, by coming in contact with a building, where there is no metallic rod to conduct it to the ground.

Sect. II. *Production of Heat and Light.*

358. Electricity, when it is retarded in its progress by a bad conductor, as in traversing the air, or when it is in large proportion to a good conductor through which it is passing, produces light and heat. An electric spark is procured on ap-

proaching the hand, or any conducting substance, to the prime conductor of the electrical machine, and a much more brilliant spark is observed on discharging the Leyden jar, as the electricity passes to the ball of the discharger.

359. If the wires at the positive and negative poles of the galvanic battery be brought in contact after charging it with diluted acid, similar appearances are presented, but the electricity cannot pass through the air to any distance, as in the preceding case; if the wires be tipt with dry charcoal, the most intense and brilliant light appears as they approach each other; and the air being heated to a great degree, the electricity still passes from the one to the other, even when they are separated 1-4th of an inch or more from each other. If extremely thin iron-wire, or leaves of gold, silver, tin, or Dutch gold, be interposed between the connecting wires, they are dissipated in vapour, and some of them are oxidated by the action of the air.

360. In experiments of this kind with the galvanic battery, a small number of very large plates is found more effectual than a great number of small plates, as it is of importance to excite as much electricity as possible within a given time. Some batteries have been constructed with plates containing many square feet of surface.

Sect. III. *Excitation of Magnetism.*

361. Professor Oersted made the interesting discovery, that the copper-wire connecting the poles of a galvanic battery affects the magnetic needle; and by the researches of Ampere, Arago, and Faraday, the connection of these forces has been distinctly proved. Iron is rendered instantly magnetic by passing electricity from a galvanic battery through a copper wire wound round it; and if a piece of iron round which a copper wire covered with silk thread has been coiled, be made to touch the poles of a horse-shoe magnet, the copper wire will be found to be electrically excited at the moment the contact is made, and capable of producing an electric spark at the instant its extremities are connected together. Recent investigations have led to the belief, that magnetism may be induced in a much wider range of substances than had hitherto been anticipated.

362. Dr Faraday, Nobili and Antinori, and Professor Forbes, were the first who obtained an electric spark from a magnet; Faraday's previous researches in electro-magnetism facilitated the discovery. Dr Ritchie, and also Pixii of Paris, have contrived different pieces of apparatus with which chemical experiments may be performed by the electric spark procured from a magnet.

363. Instruments have been invented for indicating the nature and degree of excitement produced in bodies under galvanic action, by their power in deflecting the magnetic needle; they have been termed GALVANOMETERS.

Sect. IV. *Influence of Electricity on Chemical Action.*

364. Electricity promotes combination or decomposition according to the circumstances under which it is applied. Chemical action induced by electrical arrangements, has been termed ELECTRO-CHEMICAL ACTION.

365. Electricity produced by the electrical machine, and by the galvanic battery, produce the same effect, but the latter is preferred, and almost solely used in chemical experiments, from the very large quantity of electricity which can with comparative facility be procured; electricity from the machine is extremely feeble in effecting decomposition. A battery composed of many troughs is also better suited for this purpose, though the plates do not exceed five or six inches square, than one where there is a smaller number of large plates. The effect of electricity in promoting decomposition has hitherto been more particularly studied than its power in promoting combination.

366. All bodies have a peculiar relation to electricity, so that when they are separated from their various combinations, they are attracted to the positive or negative pole. In the following table by Berzelius, the first 21 are called ELECTRO-NEGATIVES, as, when separated from other substances by electricity, they are attracted to the positive pole; the others have received the name of ELECTRO-POSITIVES, being attracted under similar circumstances to the NEGATIVE POLE. Bodies are considered to have or acquire during decomposition a different electric condition from that of the pole where they appear, and hence the terms applied to them.

Electro-Negatives.		Electro-Positives.		
Oxygen.	Tungsten.	Potassium.	Manganese.	Copper.
Sulphur.	Boron.	Sodium.	Zinc.	Silver.
Nitrogen.	Carbon.	Lithium.	Cadmium.	Mercury.
Chlorine.	Antimony.	Barium.	Iron.	Palladium.
Iodine.	Tellurium.	Strontium.	Nickel.	Platinum.
Fluorine.	Columbium.	Calcium.	Cobalt.	Rhodium.
Phosphorus.	Titanium.	Magnesium.	Cerium.	Iridium.
Selenium.	Silicium.	Glucinum.	Lead.	Gold.
Arsenic.	Osmium.	Yttrium.	Tin.	
Chromium,	Hydrogen.	Aluminum.	Bismuth.	
Molybdenum.		Zirconium.	Uranium.	

367. Again, when compounds consisting of bodies arranged in the same list are decomposed, one of them is found to be electro-positive or electro-negative in regard to the other. Thus oxygen, if previously combined with any substance in the first list, is always separated at the positive pole; and potassium, if already in combination with any other body in the second list, appears always at the negative pole. Oxygen is the most electro-negative, and potassium the most electro-positive of all known bodies.

368. In effecting electro-chemical decomposition, the body subjected to the galvanic influence is placed between the wires connecting the poles of the battery, and, however imperfectly, must conduct the electricity, otherwise no effect takes place.

369. An element or other substance, under the influence of electricity, may have its ordinary power of combination completely suspended, and be passed through bodies for which it has an affinity without combining with them.

370. The chemical agency of electricity was discovered by Nicholson and Carlisle. Sir Humphry Davy, entertaining the idea that chemical attraction depends on bodies being in different electrical conditions, and thereby attracting each other, considered that all compounds could be decomposed by the action of the electric power; proceeding on this principle, he discovered the composition of the earths and alkalis, and the power of electricity in suspending and modifying ordinary chemical attraction.

371. The influence of electricity on chemical action, has not been so clearly unfolded as to admit of its precise relation to it being very accurately defined. But Sir H. Davy's opinion that chemical and electrical action depend upon the same primary cause acting on matter, is daily gaining ground. Though numerous valuable communications have appeared since Davy's discoveries, none are equal in extent and importance to those of Dr Faraday. The following paragraphs contain a short summary of the more important circumstances which he has pointed out.

Dr Faraday's Researches.

372. Dr Faraday has confirmed by numerous experiments the views that have

been generally adopted as to the identity of the electricity of the electrical machine and of the galvanic battery; a piece of paper, moistened with a solution of iodide of potassium and starch, was found of much use as a test of electricity in his researches, the iodine being set at liberty and rendering the starch blue. The great difference in the chemical agency of the electricity of the electrical machine and the galvanic battery, is explained by a comparison of the quantity of electricity produced by each. *An electric battery of* 15 *jars, each containing* 234 *square inches of glass coated on each side, and charged by* 30 *turns of a powerful plate electrical machine,* 50 *inches in diameter, yields no more electricity than a platina and a zinc wire, each* 1–18*th of an inch in diameter, placed* 5–8*ths of an inch deep, in a diluted acid composed of* 1 *drop of sulphuric acid and* 4 *ounces of water, during* 3 *seconds.* Again, the quantity of electricity associated with the elements of bodies in their various chemical combinations is so great, *that, while electricity sufficient to decompose a grain of water can be easily procured during the chemical action of a few grains of zinc on a diluted acid, Dr Faraday has inferred that no less than* 800,000 *charges of the electric battery just mentioned would be necessary to produce the same effect.*

373. The conducting power of bodies has long been known to influence much the manner in which they are affected by electricity. Dr Faraday has added the following remarks from an extensive series of observations. Water was stated by Davy to be essential to electro-chemical decomposition; but this opinion can no longer be maintained.

374. " All bodies conduct electricity in the same manner from metals to lac and gases, but in very different degrees.

375. " Conducting power is, in some bodies, powerfully increased by heat, and in others diminished, yet without our perceiving any accompanying essential electrical difference, either in the bodies or in the changes occasioned by the electricity conducted.

376. " A numerous class of bodies insulating electricity of low intensity when solid, conduct it very freely when fluid, and are then decomposed by it.

377. " But there are many fluid bodies which do not sensibly conduct electricity of this low intensity: there are some which conduct it and are not decomposed; nor is fluidity essential to decomposition.

378. " There is but one body yet discovered which, insulating a voltaic current when solid, and conducting it when fluid, is not decomposed in the latter case.

379. " In numerous cases, as a solid becomes fluid, it loses almost entirely the power of conduction for heat, but gains in a high degree that for electricity," and *vice versa.*

380. The extremities or poles of an electric circuit are usually spoken of as having peculiar attractive and repulsive powers, to which the chemical agency of electricity has been referred; and Davy, with other chemists, considered that the decomposing agency was feebler in other parts of the circuit than at the poles. But Faraday's views are different; the force of the electric current is equal, he considers, throughout all the circuit, and that it is more correct to speak of the particles acted on by the electricity in relation to the current passing through them, than to the poles, which are merely the doors through which it passes.

381. The transferring power of electricity, by which the particles are made to appear at different poles, is much influenced by the attractive power of the substances associated with them. Thus more sulphuric acid is carried to the positive pole of a battery, if it be combined previously with soda, which attracts it strongly, than when merely combined with water, which has a weaker affinity for it.

382. Elements, and other bodies not decomposable, are not affected by an electric current unless in chemical relation to other matter.

383. No determination to a pole has been observed in cases of mere mixtures.

384. Decomposition is attributed therefore, not to the agency at the different poles, but to the electricity as it affects each particle while it passes in a current.

385. Again, among the more remarkable circumstances attending electro-chemical decomposition, the appearance of the elements at a distance from each other is to be particularly noticed. This does not arise from their being attracted to the poles, but from the compound being decomposed at each pole, the electricity producing a like effect at every part of the compound through which it passes, so that a series of combinations and decompositions take place throughout the substance acted upon between the poles. Thus if

ab ab ab ab ab ab ab

represent seven particles of the compound ab, consisting of a and b, placed between the poles of a galvanic battery, then the following arrangement is produced:

a ba ba ba ba ba ba b

This is the only explanation at all satisfactory which has been offered of the appearance of the elements of a compound at the different poles, often widely separated from each other.

386. In numerous cases, the substances that appear at the different poles, are the SECONDARY, not the PRIMARY, PRODUCTS of electro-chemical decomposition, being formed or disengaged by the primary products decomposing or combining with some of the substances with which they come in contact as they are evolved.

387. Metals which have been connected with the positive pole of the galvanic battery, have been observed, even when entirely removed from it, to have the power of promoting combination: this was tried more particularly in reference to the action of oxygen with hydrogen gas. On a careful examination of the causes of this supposed electric action, it was found to depend on the surface being thoroughly cleaned in the liquid in which they were immersed, and that similar effects were produced when platinum, the metal preferred in these experiments, was carefully cleaned by sulphuric acid and water.

388. Bodies directly decomposable by electricity have generally been found to consist of 1 equivalent of each element; and in simple binary compounds this has always been observed to be the case.

389. " The chemical power of a current of electricity is in direct proportion to the absolute quantity of electricity which passes."

390. The electricity which decomposes, and that which is evolved by the decomposition of a certain quantity of matter, are alike.

391. The quantity of electricity required for the decomposition of an equivalent weight of a binary compound, consisting of 1 equivalent of each element, is the same with that necessary to effect the decomposition of an equivalent quantity of any similar binary compound.

392. The term ELECTRO-CHEMICAL EQUIVALENTS refers to the quantities of substances separated from any compound by a given amount of electricity; and referred to a standard of comparison, they are found to approximate as nearly as could be expected to the ordinary equivalent numbers.

393. The definite nature of electro-chemical action, and the correspondence between the chemical equivalents of bodies ascertained in the usual manner, and from the quantity separated by a given amount of electricity, are equally novel and important in the history of chemical science.

394. In pursuing his researches, Dr Faraday found it necessary to employ an instrument, which he contrived with the view of measuring the quantity of electricity produced, and employed in different experiments. Having shewn that the chemical decomposing action of a current of electricity is constant for a constant

quantity, however variable in its course, in its intensity, and in the nature of the poles and conductors employed, water was chosen to indicate the strength of the current passing through it by the quantity decomposed, being acidulated with sulphuric acid to make it a better conductor. The apparatus employed consists of a vessel containing the fluid, in which two wires or plates of platinum are introduced, placed at a little distance from each other. The wires being connected with the galvanic battery, so as to form part of the circuit, the quantity of water decomposed at the same time that other effects are produced by passing the electricity through different substances in another part of the circuit, affords an excellent standard of comparison. The instrument is termed the VOLTA ELECTROMETER.

395. Dr Faraday has introduced the following new terms, to save circumlocution, and express more specifically than he could with those now in use many of the circumstances attending electro-chemical decomposition.

396. ELECTRODES, from ἤλεκτρον and ὁδος, a way, used instead of poles, as they are regarded merely as the doors by which electricity passes; they are the boundaries of the decomposing matter in the direction of the electric current.

397. ANODE, ἀνα, upwards, and ὁδος, a way; that part of the surface of a decomposing body which the electric current enters; the part immediately touching the positive pole.

398. CATHODE, κατα, downwards, and ὁδος; that part of the decomposing body which the current leaves; the part next to the negative pole.

399. ELECTROLYTES, ἤλεκτρον and λυω, I separate; a body which can be decomposed directly by electricity. ELECTROLYZED means electro-chemically decomposed.

400. ANION, ἀνιον, that which goes up; a body which passes to the positive pole, to the anode of the decomposing body, as it is separated by electricity.

401. CATION, κατιον, that which goes down; a body that passes to the negative pole, to the cathode of the decomposing body. IONS signify anions or cations.

SECT. VI. *Effect of Electricity on Animals.*

402. Electricity has a powerful effect upon the animal economy, producing instant death when in great intensity, as a flash of lightning, or acting merely as a gentle stimulus, when applied in minute quantity.

403. If any individual stand upon a stool supported on glass feet, that he may be insulated by a non-conductor, and touch the prime conductor of the electrical machine, electricity will then accumulate in him, and a spark may be obtained from him by any conducting substance, in the same manner as from the prime conductor itself.

404. When an electric shock is to be communicated to any part of the body, it must be made the medium of communication between the positive and negative poles of the battery—it must form part of the circuit through which the electricity passes. Metallic rods, brass chains, and the common electric discharger, are of much use in making the necessary connections.

405. Galvanic electricity having much less tension than that produced by the electric machine, will not pass through the dry cuticle; it must be moistened with a saline solution, or very dilute acid.

406. When it is necessary to pass a strong galvanic current through any part of the body, it may be often advantageous, as Mr Kemp suggested, to avoid the shock which is usually communicated on making or breaking the circuit. For this purpose, the connection having been made between one pole of the battery and one extremity of the part through which the electricity is to be passed, Mr

Kemp recommends a metallic plate or ball, terminating in two wires, to be employed in completing it. The plate or ball being placed on the other extremity of the part alluded to, one of the terminating wires is made to touch the pair of plates next to the pole of the battery first connected; while this connection is maintained, the other wire is moved beyond it, and then the first lifted past it again, taking care always to maintain the connection with one wire while the other is moved forward to include an additional number of plates. In this manner, not only is the first shock avoided, but also the second, if care merely be taken that the wires shall be moved back in the same manner as they were advanced.

407. Electricity acting so powerfully on the system, is used with advantage in numerous diseases of a paralytic nature, and also in cases of suspended animation.

408. Even for some time after death, the muscles are powerfully affected by it, and thrown into the most violent convulsive movements.

* Professor Daniell has lately constructed a very ingenious and useful galvanic battery, in which, by a series of syphons, the charge can be continually renewed without disturbing the apparatus, a small quantity of fluid dropping constantly into each syphon, and displacing an equal bulk of the former charge. He has termed it the CONSTANT BATTERY.

PART II.

CHEMICAL HISTORY OF INDIVIDUAL SUBSTANCES.

DIVISION I. INORGANIC BODIES.

CLASS I. NON-METALLIC SUBSTANCES.

CHAP. I. OXYGEN.

409. OXYGEN, from ὀξυς, sour or acid, and γεννάω, I produce. When this name was given, all acids were believed to contain this element: numerous exceptions have since occurred. *Old Names.*—Vital, Empyreal, and Dephlogisticated Air.

410. Symb. O, or a dot repeated as often as there are atoms of oxygen.—Eq. 8 by W, and ▭ by V.—Sp. gr. 1.111.—W. of 100 c. i. 33,888 grs.—Water, 100 c. i. absorb 3.7 of oxygen.

411. Gaseous, transparent, colourless, inodorous, uninflammable, supports combustion brilliantly; its compounds, when not *acid*, are termed OXIDES; the ingredient combining with oxygen is said to be OXIDIZED, or OXIDATED; the process is called OXIDATION. DEOXIDATION signifies the separation of oxygen.

412. Oxygen is the most abundant of the elements, forming upwards of one-half of the globe; it exists in air, water, most earthy substances, and in almost all the products of the vegetable and animal kingdom.

413. The principal OXIDATING AGENTS are air, water, oxacids, and salts containing them, as nitrates and chlorates. The principal DEOXIDATING AGENTS most generally used are carbon, hydrogen, phosphorus, and potassium; other inflammable metals are also employed.

414. Prepared commonly from oxides, or salts of nitric or chloric acid. Peroxide of manganese, peroxide of mercury, nitrate of potassa, and chlorate of potassa are the compounds generally used.

415. PEROXIDE OF MANGANESE 44, = oxygen 16 + manganese 28; 40 of sulphuric acid (49 aqueous acid are used) discharge 8 of oxygen by heat. Sulphate of manganese, consisting of the acid and 36 oxide, remain in the glass retort used.

Fig. 2.

Materials.			Products.
44 Peroxide,	Oxygen	8	8 Oxygen.
	Protoxide	36	
40 Sulphuric Acid,.........		40	76 Sulphate of Manganese.

:Mn&⁝S = ⁝S·Mn&O.

Carbonic acid, from carbonate of lime mixed with the peroxide, is often evolved at first from bad manganese, and a burning taper is extinguished by this gas, but burns brilliantly when the oxygen is pure. An excess of acid facilitates the expulsion of the oxygen.

416. For large quantities of oxygen, the peroxide alone is heated to redness in iron bottles: 88 (44 × 2) give 8 of oxygen, deutoxide of manganese remaining = oxide 36 + peroxide 44 = 80.

Fig. 3.

	Materials.		Products
44 Peroxide	{ Oxygen...	8	8 Oxygen.
	{ Protoxide	36	
44 Peroxide	....	44	80 Deutoxide.

2:Mn = ⁝Mn²&O.

417. PEROXIDE OF MERCURY, 216 = oxygen 16 + mercury 200. All the oxygen is detached by heating it in an iron bottle. The mercury is distilled. :Hg = Hg&O².

418. NITRATE OF POTASSA = acid 54 + potassa 48. Heated in an earthen or iron bottle, part of the oxygen is expelled after it is melted; compounds of potassa with various proportions of oxygen and nitrogen remain; their precise composition is uncertain. By long continued heat potassa alone is left, corroding in general the vessel in which it is heated. The first gas evolved is the purest; much nitrogen is associated with the oxygen ultimately disengaged.

419. CHLORATE OF POTASSA = chloric acid 76 + potassa 48 = chlorine 36 + potassium 40 + oxygen 48. :⁝Cl·K = ClK&O⁶. By heat all the oxygen is expelled both from the chloric acid and the potassa: chloride of potassium 76, remain. This salt gives the purest oxygen.

420. If oxygen gas be perfectly pure, 8 by W, = ▭ by V, is entirely condensed by hydrogen 1 by W, = □ by V, on the application of heat or spongy platinum. See Hydrogen, 439.

421. Oxygen is largely consumed in numerous natural and artificial operations, as respiration, combustion, vegetation, and the preparation of oxides, many acids, salts, and other compounds. There are few processes to which it does not contribute,—during combustion communicating the heat required, entering into the new combination, or otherwise taking a part in the action.

422. COMBUSTION, or the evolution of heat and light during chemical action, very frequently attends the combination of oxygen, which has hence been called a SUPPORTER OF COMBUSTION, a term now used in a very general sense, and applied to any substance with which another can evolve heat and light by chemical action. At one time it was believed that no inflammable could burn without oxygen. This opinion is no longer tenable, but in all ordinary cases, as in the combustion of coal, coal-gas, oil, and other common inflammables, the oxygen of the air combines with the inflammable body. Formerly Stahl imagined that all combustible bodies contained a peculiar matter, which he termed PHLOGISTON, the phenomena of combustion being produced as it is disengaged. He overlooked the influence of the air, which Lavoisier pointed out, who proved that it was in part consumed, that the inflammable gained as much as the air lost, that oxygen was the ingredient removed from the air by the burning body, and that it resumed its original character, and was again inflammable when the oxygen was separated from it.

423. Many inflammables, as phosphorus, coal, and oils, can combine with oxygen at a less elevated temperature than is necessary for a rapid combustion. This SLOW COMBUSTION is often attended with the evolution of light visible in the dark only; and when the air has sufficiently free access, and is not carried too ra-

pidly away, it often terminates in ordinary combustion. Numerous accidents are attributed to this source.

424. The source of the heat and light produced by combustion during the new arrangement of the particles of the combining substances, has not been satisfactorily explained.

CHAP. II. HYDROGEN.

425. HYDROGEN, from υδορ, water, and γεννaω. *Old name.*—Inflammable Air.

426. Eq. 8 by W, □ by V.—Sp. gr. 0.0694.—W. of 100 c. i. 2.116 grs.—Water 75 c. i. dissolve about 1 of hydrogen.

427. Gaseous, transparent, colourless, the lightest of the elements, inflammable, does not support combustion nor respiration. During combustion, hydrogen 1 = □, combines with oxygen 8 = ▭, and forms water 9 = □, estimating it in the state of vapour.

428. Hydrogen burns in contact with oxygen only when heated; if mixed previously with the oxygen required, or with air containing as much oxygen, ⊟⊟▭ measures, it takes fire with explosion on the approach of flame: an excess of either ingredient diminishes the explosive violence; it is most violent with the pure gases. The electric spark, sudden and powerful compression, several metals in a minute state of division, more especially spongy platinum, or a plate of thin metal extremely clean, explode the mixed gases. At a less elevated temperature than that produced by flame, with the gaseous mixture rarefied by a diminution of pressure, or by diluting it with either of the gases in excess, there is no explosion, a slow combustion ensues.

429. Doebereiner discovered the power of platinum in a minute state of division of inflaming hydrogen mixed with air; the platinum absorbs a large quantity of gas, heat is evolved, the metal becomes incandescent, the gas is inflamed.

430. During combustion a pale looking flame appears, but if an explosive mixture be compressed to half its volume, and then inflamed, a brilliant light is seen. Much heat is evolved as hydrogen burns: By the combustion of 1 of hydrogen 320 of ice can be melted. (Dalton.)

431. In the OXYHYDROGEN BLOWPIPE, a stream of hydrogen being supplied with pure oxygen as it escapes from the nozzle of the apparatus, a most intense heat is maintained with great facility. Dr Hare contrived this powerful instrument: modifications have since been proposed by Newman, Wollaston, Gurney, and others. In these the gases are mixed in the same vessel. In that by Hemming, which is preferred, instead of passing the gas through oil, water, or wire-gause, it escapes through a safety tube, 6 inches long, filled with fine brass-wire in length, packed as closely as possible, and fixed in their position by forcing amongst them a pointed metallic rod. The gases escape between the wires in the tube. Such precautions are necessary to prevent explosion from the return of the flame into the reservoir. When the instrument is made on a large scale the gases are not mixed, but kept in separate vessels until they are introduced into the nozzle, from which they escape.

432. Professor Daniell has lately recommended a new jet, where a stream of oxygen is introduced into the interior of a stream of hydrogen; he has found it very useful.

433. Hydrogen, though extensively DISTRIBUTED through the globe, is not nearly so abundant as oxygen; it forms 1-9th of the water of the globe, and is found in almost all the products of the animal and vegetable kingdoms. Water is the great source of hydrogen for experimental purposes.

434. Hydrogen is always PREPARED from water when required in a state of purity. If watery vapour be transmitted over iron-turnings heated to redness, iron 28 decompose water 9; 1 of hydrogen is disengaged, and 36 oxide of iron formed as the iron unites with oxygen, the other element of the water. Fe&·H = ·Fe&H.

435. Hydrogen is more conveniently procured by mixing iron or zinc with water and sulphuric acid; 40 of acid + iron 28 + water 9, give 1 of hydrogen and 76 of sulphate of iron, the latter consisting of 40 acid and 36 oxidated metal.

Fig. 4.

Materials.			Products.
9 Water	Hydrogen	1	1 Hydrogen.
	Oxygen...	8	
28 Iron................		28	
40 Sulphuric Acid ...		40	76 Sulphate of Iron.

Fe&·H&⁝S = ⁝S·Fe&H.

Much more water must be present than is necessary for decomposition, otherwise each little mass of metal is soon covered with a crust of sulphate, and the process stops. The additional water dissolving the sulphate, as it is produced, leaves the metal brilliant and ready to decompose more water. About 7 or 8 parts of water by bulk to 1 of acid are generally used.

436. Zinc 34 produce the same effect as 28 of iron, decomposing 9 of water, taking 8 of oxygen and separating 1 of hydrogen. The same quantity of acid is also required, and a corresponding quantity of sulphate of zinc produced, consisting of 42 oxide + 40 acid = 82. ⁝S&·H&Z = ⁝S·Z&H.

337. In both cases a small portion of black matter is separated from the metal, and floats undissolved in the excess of fluid: charcoal is the principal ingredient from the iron. During the preparation of the gas by this process, considerable heat is produced, both during the mixing of the acid with the water, and the solution of the metal. The solutions of sulphate of iron and sulphate of zinc give crystals on cooling when they are concentrated.

438. The hydrogen, as at first prepared by these processes, and more especially by the iron, has often a disagreeable smell, from the presence of minute quantities of sulphur, carbon, or other substances.

439. PURE HYDROGEN condenses exactly half its bulk of oxygen, when detonated with it by the electric spark, or subjected to the action of spongy platinum. In conducting this operation, a given bulk of hydrogen is introduced into a tube filled with mercury, and resting on the mercurial trough; an excess of oxygen being added to it, the platinum ball is introduced; an excess of one of the gases is necessary to prevent explosion. Two-thirds of the amount of condensation indicate the bulk of pure hydrogen.

440. Hydrogen is much employed in numerous chemical experiments, especially as a deoxidating agent, and for the production of intense heat in the oxyhydrogen blowpipe. From its great levity it is the best material for filling balloons.

SECT. I. *Water or Oxide of Hydrogen.*

441. Symb. ·H = oxygen 8 + 1 hydrogen.—Eq. 9 by W, □ by V.—Sp. gr. 1, being the standard of comparison in estimating the specific gravities of solids and liquids. About 828 times heavier than air; W. of 100 c. i. at 62, bar. 30 = 252.458 grs., boils at 212°, barometer being 30, expanding to 1696 times the volume it occupies at 39.38°. Compounds termed HYDRATES OR AQUEOUS COMBINATIONS.—For its relations to caloric see 150, 152, 184, &c. Refracts light powerfully; conducts electricity imperfectly.

442. Transparent, colourless, inodorous, tasteless, compressible by great pressure.

443. As a chemical agent, distinguished by extensive range of combination, power of solution, being the medium of combinations and decompositions, so numerous, as to be connected with the greater number of the phenomena of nature and art. In these a part is frequently decomposed, its elements forming new compounds; it is, on the other hand, formed by numerous compounds as they decompose each other.

444. Metallic oxides and compounds of hydrogen, when they decompose each other, almost invariably produce water and a new metallic compound, as when sulphureted hydrogen and oxide of lead form sulphuret of lead and water. Again, metallic compounds, not oxides, when they decompose water, generally produce a metallic oxide and a new compound of hydrogen, as when sulphuret of iron decomposes water, forming oxide of iron and sulphureted hydrogen.

445. Water absorbs a number of gases. The following Table has been drawn up principally from the experiments of Dalton, Henry, and Saussure:—

Gases.	Water 100 c. i. absorb	Gases.	Water 100 c. i. absorb
Sulphureted hydrogen	253 vols.	Oxygen . .	3.7 vols.
Carbonic oxide .	100	Hydrogen . .	1.56 ?
Nitrous oxide . .	76	Nitrogen . .	1.56 ?
Olefiant gas . .	12.5	Carbonic oxide .	1.56 ?

The numbers representing the absorption of the last three gases are considered by some chemists rather low: Saussure made them as high as 4 and 6, but his estimate was too high. In making experiments of this kind, recently boiled water must be employed.

446. The quantity of gas absorbed by water is, according to Dr Henry's experiments, proportional to the pressure. The bulk of gases being also proportional to the pressure, the law may be stated in these terms:—*Water always absorbs the same* BULK *of gas, whether it is subject to the usual or to an increased pressure;* for example, the pressure being doubled, the weight of gas absorbed is also doubled; but, previously to this, the pressure applied will have diminished its bulk to one-half, so that the *volume* absorbed is exactly the same as at the ordinary pressure.

447. Water absorbs oxygen and nitrogen from the air, condensing more oxygen than nitrogen. Dalton has maintained that water absorbs the same quantity of gases in a mixed state, which it would do were they separate.

448. Water charged with air becomes turbid when salts are dissolved in it, the air escaping in very minute globules, which rise slowly to the top. All common water contains air.

449. Water is also found in all air obtained at the surface of the earth. It is separated most effectually from any given quantity of air, by including it in a jar over a mercurial trough with chloride of calcium. Sulphuric acid is much used for the same purpose.

450. Water is never met with absolutely pure in nature, being always charged more or less with impregnations from the air, or various animal, vegetable, and mineral substances, with which it may have come in contact.

451. PURE WATER for chemical solutions, tests, &c. must be prepared by distillation; the first portions are rejected, being frequently mixed with air. Glass vessels are necessary for distillation only in very delicate experimental investigations. Distilled water, which does not redden litmus, gives no precipitate with muriate of baryta, nitrate of silver, or oxalate of ammonia, may be considered fit for all ordinary purposes where distilled water is required, as these tests would

detect any of those matters usually dissolved in common water, should the distillation have been conducted in an imperfect manner.

452. RAIN-WATER and SNOW collected at a distance from town, and where they cannot be exposed either to dust or vapours which they might absorb, are the purest natural waters. Even these, however, must be boiled, to expel air and carbonic acid, before they can be substituted for distilled water in various experiments; and occasionally, more especially after lightning, they have been found to contain minute quantities of nitric acid.

453. SPRING WATER having flowed in contact with various rocks and minerals, more or less soluble, contains in general earthy salts in solution, in addition to a variable proportion of gases. These give it a more grateful taste than distilled water, which is very insipid; when the impregnation of foreign matter is so great as to prevent it from being used for domestic purposes, it is termed a MINERAL WATER. RIVER WATER belongs to the same class as spring water. WELL WATER, and the water of lakes, and ponds is generally more contaminated with foreign matter; while in marshes it is loaded with putrescent vegetable remains.

454. HARD WATER owes its peculiar properties, in general, to earthy salts. Some kinds are much improved by boiling, as those charged with lime retained in solution by excess of carbonic acid; the heat expels the excess, and the lime with the rest of the carbonic acid is precipitated. On other occasions the addition of carbonate of soda or carbonate of potassa is advantageous, the alkali (soda or potassa) uniting with any acid matter forming part of the saline ingredients of the water, and preventing them from decomposing soap and other substances.

455. When water has acquired an offensive odour, this may generally be removed by filtering it through sand and charcoal. Casks charred in the inside are frequently employed for preserving water on long voyages. IRON TANKS have been lately used for the same purpose. Leaden vessels are dangerous, being frequently much corroded, and communicating an impregnation to the water.—See Lead.

SECT. II. *Binoxide of Hydrogen.*

456. Symb. :H = oxygen 16 + hydrogen 1.—Eq. by W. 17.—Sp. gr. 1.462.—Synon. Peroxide of hydrogen, oxygenated water.

557. Liquid, transparent, colourless, inodorous. Slowly decomposed by light; effervesces from escape of oxygen, at the temperature of 59°, water being left, and with explosion at 212°. Many metals in a minute state of division, and metallic oxides, produce the same effect; but some metals, and oxides having a great attraction for oxygen, as potassium and oxide of tin, combine with this element as it is separated from the water. Oxides which are reduced by a red heat are also reduced in contact with it, as those of gold, platinum, and silver.

458. The addition of water and some acids give more permanency to this compound, and acid compounds were formerly described, similar in many respects to oxygenated water, but they are now regarded as compounds of oxygenated water with the acid.

459. Thenard prepared this interesting compound in the following manner:—Baryta is charged with oxygen till binoxide of barium is formed, which is dissolved by muriatic acid, diluted with 10 parts of water. By sulphuric acid the baryta is precipitated as sulphate of baryta, the oxygen combining with a portion of the water, and forming binoxide of hydrogen, which remains with the rest of the water and muriatic acid. This is repeated till the water shall have taken up about 25 times its bulk of oxygen. On adding sulphate of silver to the liquid (which

contains water, binoxide of hydrogen, and muriatic acid), the oxide of silver in the sulphate removes the muriatic acid, which is replaced by the sulphuric acid of the sulphate. Baryta is then added to precipitate the sulphuric acid; and the water still associated with the binoxide, is removed by evaporation in an exhausted receiver, a plate with sulphuric acid having been introduced to condense the watery vapour. The minute details necessary to insure the success of the process are given by Thenard in his Traité de Chimie.

CHAP. III. NITROGEN.

460. Name derived from nitrates, compounds of nitric acid, of which it is an element.—Synon. Azote, from α privative, and ζωη, life.

461. Symb. N.—Eq. 14 by W. and □ by V.—Sp. gr. .9722.—W. of 100 c. i. 29.652 grs.—Water 100 c. i. absorb about 2.

462. Gaseous, transparent, colourless, inodorous, uninflammable, incapable of supporting combustion or respiration.

463. Berzelius has entertained the idea that nitrogen is a compound, but no experimental evidence has supported his opinion. Mercury subjected to galvanism in compounds containing nitrogen, as the muriate of ammonia, increases in size, apparently by the addition of metallic matter; but minute examination has proved that the increased bulk depends on its mechanical admixture with ammonia and hydrogen.

464. Nitrogen forms 4-5ths of atmospheric air by bulk; it is found in two or three mineral substances, as coal and nitrates, in few vegetable principles, but abounds in the products of the animal kingdom.

465. Prepared commonly by burning phosphorus in a jar of air, placed over water. Air = nitrogen 36 + oxygen 8; the phosphorus combining with the oxygen forms an acid, which the water removes. The nitrogen retains a very minute portion of the phosphorus, which may be removed, if required, by a solution of potassa.

466. A mixture of sulphur 1 + iron-filings 2, when moistened and gently heated, and water in which lime and sulphur have been boiled, are also used to remove oxygen from air. Nitrogen may be prepared from fibrine by diluted nitric acid. —See Fibrine.

Sect. I. *Oxide of Nitrogen.*

467. Symb. ·N = nitrogen 14 + oxygen 8.—Eq. 22 by W, □ by V.—Sp. gr. 1.527.—Water 100 c. i. absorb 76.—Condensed into a liquid by a pressure of 50 atmospheres at temp. 45°.—Resolved by a bright red heat in a porcelain tube into oxygen, nitrogen, and nitrous acid.—Synon. Protoxide of Nitrogen, Nitrous Oxide.

468. Gaseous, transparent, colourless, odour slight, taste sweet, supports combustion much more brilliantly than air; sulphur burns in it with a distinct rose tint. It detonates with an equal bulk of hydrogen. In these experiments the nitrogen is separated; occasionally binoxide of nitrogen and nitrous acid are produced.

469. Acts powerfully on the animal economy when inhaled from a bag, as Sir H. Davy discovered. Much excitement is produced, accompanied by a rapid succession of vivid ideas; a rather agreeable sensation is perceived in the chest, followed by an irresistible propensity to laughter and muscular exertion. The excitation is so great, when much is taken, that assistants should always be ready in

case of accident. The sensations it produces are highly pleasing in general; occasionally the reverse is observed; headach, and even fainting, are sometimes produced. The effects are very transient, seldom continuing in full force above from 1 to 3 minutes. They are said to be more easily produced each succeeding trial; this depends, in many instances at least, on its being inspired in a more perfect manner. Another peculiarity is, that the excitement is not accompanied by any corresponding depression, when there has not been great muscular exertion.

470. Prepared from nitrate of ammonia by heat. 71 of this salt = 54 nitric acid + 17 ammonia. But 54 acid = nitrogen 14 + oxygen 40, and 17 ammonia = nitrogen 14 + hydrogen 3, therefore 71 nitrate of ammonia = 28 nitrogen + 3 hydrogen + 40 oxygen = 44 oxide of nitrogen and 27 water, or 2 atoms of oxide of nitrogen and 3 of water.

Fig. 5.

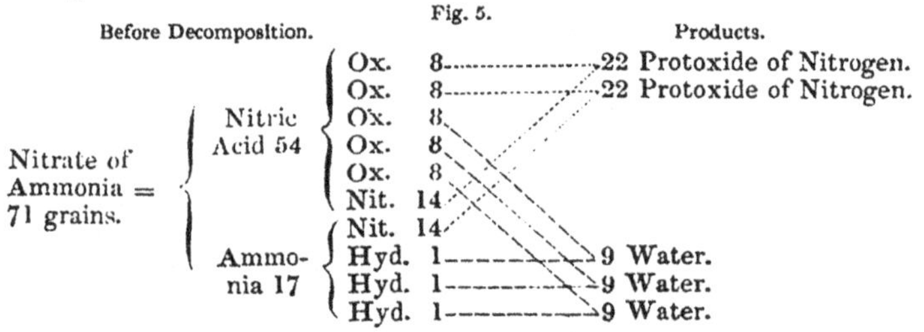

Symb. ⁝N+NH^3 = 2·N&3·H.

471. Nitrous oxide is not obtained from the nitrate of ammonia till it has been freed in a great measure from the water of crystallization by fusion, and heated to a temperature between 400° and 500°.

Sect. II. *Binoxide of Nitrogen.*

472. Symb. :N = nitrogen 14 + oxygen 16.—Eq. 30 by W, □□ by V.—Sp. gr. 1.0416.—W of 100 c. i. 31.77 grs.—Synon. Deutoxide of Nitrogen, Nitrous Gas, Nitric Oxide.—Decomposed by a red heat in porcelain tubes.

473. Gaseous, transparent, colourless; it cannot be respired, producing ruddy fumes of nitrous acid when mixed with the air, by combining with its oxygen; with pure oxygen, still more dense vapours are seen. Few inflammables burn in it, but phosphorus, at an elevated temperature, burns with great brilliancy. With an equal bulk of hydrogen, a silent combustion ensues on the application of heat, and a bluish flame appears. Solutions of sulphate and muriate of iron absorb it rapidly, and become dark.

474. Nitric, nitrous, and hyponitrous acids, may all be formed by the action of this gas on air or oxygen; the nature of the products is influenced by the size of the vessel in which they are mixed, the presence of moisture, the proportion of the gases, and numerous other circumstances. When the experiment is made in jars over water, all the acid matter is immediately condensed. According to Gay Lussac, when the binoxide is added in excess to the air, or other gaseous mixtures containing oxygen in a wide vessel over water, the same products are always formed, and the amount of condensation, divided by 4, expresses the quantity of oxygen in the air or mixture under examination.

475. Prepared from aqueous nitric acid, diluted with from 2 to 3 times its bulk of water, by the action of copper or mercury, heat being applied, when the gas is disengaged very slowly. Nitric acid 54 = nitrogen 14 + oxygen 40; 16 of oxygen being left with the 14 of nitrogen produce the gas, the remaining

oxygen combining with the metal. Each eq. of metal takes 1 eq. of oxygen, so that 3 eqs. of metal are required for 1 eq. of acid decomposed. Farther, 3 eqs. of acid must be employed which are not decomposed, combining with the oxidated metal, and forming a salt, which remains in solution. 4:⁝N&3Cu = :N& 3(:⁝N·Cu.)

476. Nitric oxide is produced in numerous chemical operations when nitric acid is decomposed; it was employed much at one time in eudiometrical experiments.

Sect. III. *Hyponitrous Acid.*

477. Symb. ⁝N = nitrogen 14 + oxygen 24.—Eq. 38 by W.—Formed by cooling a mixture of oxygen and nitrogen gases. Its existence was inferred from the action of a solution of potassa on nitric oxide mixed with oxygen in the requisite proportions, when they are entirely condensed. If an acid be added to separate the potassa, the hyponitrous acid is immediately decomposed, one Eq. gains oxygen from another, and is converted into nitrous acid: that which loses oxygen becomes nitric oxide. Liquid, green, very volatile, colourless, at 0°.

Sect. IV. *Nitrous Acid.*

478. Symb. ⁞N = nitrogen 14 + oxygen 32.—Eq. by W, 46.—Sp. gr. of vapour 3.19.—W. of 100 c. i. 97.417 grs.—Rapidly absorbed by water. Liquid, orange-yellow colour, sp. gr. 1.452, volatile, boils at 82°, decomposed by a red heat; acid taste, pungent odour; vapour transparent, deep orange-red colour; supports combustion brilliantly.

479. Nitrous acid poured into water is immediately decomposed, nitric oxide is disengaged, and the fluid acquires a greenish colour; on continuing the addition, each successive portion undergoes decomposition less and less, till at last no farther change takes place; binoxide of nitrogen is disengaged during these changes, while nitric and hyponitrous acids are also produced. The nitrous acid becomes olive-green, bluish-green, and ultimately transparent and colourless by the addition of successive portions of water.

480. Nitrous acid is much used as an oxidating agent, more especially when mixed with nitric acid, being often prepared along with it.

481. Prepared in the liquid form by distillation from nitrate of lead; the nitric acid is decomposed, part of its oxygen left with the lead; the nitrous acid vapour must be condensed in a receiver kept very cold.

482. Prepared in the gaseous form, by mixing 2 measures of binoxide with 1 of oxygen in an exhausted receiver; jars over water or mercury cannot be used: water promotes the production of nitric acid, and mercury takes oxygen from nitrous acid.

Sect. V. *Nitric Acid.*

483. Symb. :⁝N = nitrogen 14 + oxygen 40.—Name of compounds, Nitrates.—Has not hitherto been procured in an insulated form; the various compounds usually referred to under this title being combinations of the acid with water.

484. Symb. of aqueous nitric acid, 2·H+:⁝N = water 18 + acid 54 = 72.—Sp. gr. 1.48.—Transparent and colourless when pure; often straw or orange-coloured, from nitrous acid in combination; fuming; very acid and corrosive; boils below 240°. Decomposed when passed in vapour through a red-hot tube, being resolved into oxygen and nitrogen, or compounds of these gases. Light resolves a portion slowly into nitrous acid and oxygen.

485. When coloured by nitrous acid, heat expels it, and the acid becomes colourless.

486. The specific gravity of nitric acid varies according to the water combined with it. When 1.5, it contains 79.7 per cent. of acid; and Dr Thomson has calculated that, with a specific gravity of 1.55, it would consist of 54 acid + 9 water; acid of this strength is procured with much difficulty. The following Table by Dr Thomson, shews the specific gravity of acid of different strengths:—

Specific gravity.	Acid in 100 parts.	Atoms of Acid.	Atoms of Water.	Specific gravity.	Acid in 100 parts.	Atoms of Acid.	Atoms of Water.
1.4856	75.000	1	2	1.3032	40.000	1	9
1.4546	66.668	1	3	1.2844	37.500	1	10
1.4237	60.000	1	4	1.2656	35.294	1	11
1.3928	54.545	1	5	1.2495	32.574	1	12
1.3692	50.000	1	6	1.2334	31.579	1	13
1.3456	46.260	1	7	1.2173	30.000	1	14
1.3220	42.857	1	8	1.2012	28.571	1	15

487. Acid of sp. gr. 1.42 boils at 248°, rises in vapour without change, and congeals at —41°; stronger becomes weaker, and weaker becomes stronger, by boiling.

488. Nitric acid combines in all proportions with water, producing heat. Poured upon snow it melts it rapidly, producing great cold.

489. Nitric acid is a powerful oxidating agent, communicating large quantities of this element in numerous cases with great facility. Inflammables, most metals, and numerous animal and vegetable matters, act rapidly with it, more especially if heated; nitrous acid, nitrous and nitric oxides, and nitrogen, are often disengaged in various proportions, sometimes with explosive rapidity. Combustion is also observed at times when the strong ruddy acid is poured upon inflammables.

490. Nitric oxide in contact with colourless nitric acid is quickly absorbed, producing successively a straw, orange, olive, green, and greenish-blue colour. Nitrous and hyponitrous acids are produced as the acid is deoxidated.

491. PREPARATION.—Nitrate of potassa or nitre, a native compound containing nitric acid and potassa, is generally used, along with aqueous sulphuric acid, = water 9 + acid 40. Aqueous sulphuric acid 98 (2 eqs.) decompose completely, nitre 102, = 54 nitric acid + potassa 48. The nitric acid is distilled from a glass retort, and condenses in the receiver with the greater portion of the water; a small portion remaining in the retort with the bisulphate of potassa, formed by the combination of the sulphuric acid and potassa.

Fig. 6.

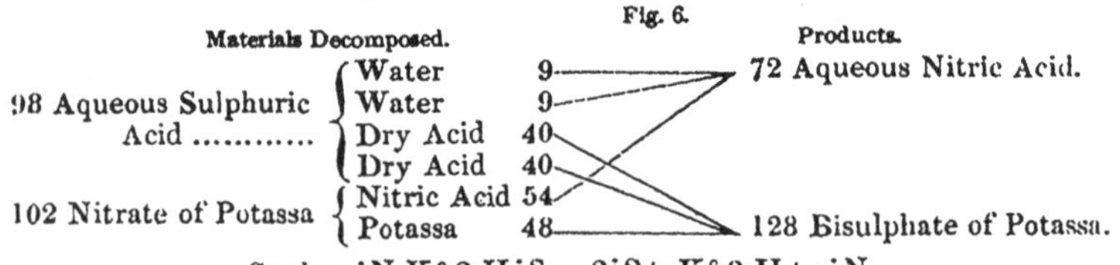

Symb. : ⁝N·K&2·H⁝S = 2⁝S+·K&2·H+: ⁝N.

492. Less sulphuric acid is sufficient to expel the nitric acid, but a higher temperature being then required, a portion is apt to be resolved into nitrous acid and oxygen. The London and Dublin Colleges recommend nearly the same proportions, viz. equal or nearly equal weights of aqueous acid and nitre. According to the Edinburgh College, less acid is used, two parts only being taken with three of nitre; in this case, the saline matter in the retort consists of sulphate and bisulphate of potassa; much nitrous is mixed with the nitric acid.

493. The AQUA FORTIS of commerce is prepared from nitre and diluted sulphuric acid and is less concentrated.

494. The diluted acid of the
Edinburgh College consists of equal weights of their acid and water,
London .... 1 measure acid + 9 water,
Dublin .... 3 .... + 4 ......

495. Minute quantities of nitric acid have been formed by passing the electric spark through the air.

496. Tests.—The best test is morphia, as proposed by Dr O'Shaughnessy, which produces a yellowish tint, succeeded by a bright orange-red colour, passing ultimately to a pale yellow. Zinc and many other metals, as copper and mercury, produce reddish fumes on adding a little water. When in the form of a nitrate, a little sulphuric acid must be added to detach the base before the tests will act. Gold-leaf mixed with muriatic acid has been much used as a test, the gold being dissolved on adding a little nitric acid; but chloric and bromic acids producing the same effect, it cannot be relied on.

497. Contaminated with sulphuric acid, baryta gives a white precipitate of sulphate of baryta. Chlorine is detected by nitrate of silver, chloride of silver being precipitated. No precipitate occurs in either case, if the acid be pure and diluted previously with water. If the precaution of diluting the strong acid be omitted, it will take the water holding the test in solution, and precipitate the test itself, even though no impurity be present.

498. Nitric acid is much employed in oxidating processes, in the preparation of many acids, aqua regia, and as an escharotic. In some cases it acts feebly, unless diluted with water. Diluted, it is used medicinally.

499. Nitric acid presents, in a very characteristic form, the properties by which acids are distinguished. It has a sour taste, is very acrid and corrosive, neutralizes salifiable bases, forming numerous salts. These compounds or nitrates are soluble in water, crystallizable, have a cool taste, and deflagrate with inflammable matter.

Sect. VI. *Atmospheric Air.*

500. Symb. $\cdot N^2$ = nitrogen 28 + oxygen 8.—Eq. 36 by W, □□▭ by V.—Sp. gr. 1, being taken as a standard of comparison for gases.—W. of 100 c. i. 30.5 grs.* at 60° F. and 30 Bar. It is nearly 828 times lighter than water. Constitution discovered by Scheele and Lavoisier.

501. Besides oxygen and nitrogen, variable quantities of watery vapour are found in the air. In hot climates, it amounts at times to 2 per cent. in weight, and varies from that to 1-300th part in cold countries. Carbonic acid is present in very minute quantities, the mean, according to Saussure, is 1 part in 2000 of air, an estimate below those generally given; at the surface of the ocean it is said to be wanting. Numerous other substances must be present at different times in the atmosphere, as odoriferous matters and carbureted hydrogen, but in too small proportion to be detected. Muriatic acid has been detected, it is affirmed, in air at the sea-shore.

502. The atmosphere is believed to extend to a height of 45 miles, diminishing in density as the elevation increases, and in temperature by 1 degree for every 300 feet it recedes from the surface.

503. Before the great uniformity in the composition of the atmosphere at all heights and in all situations in the globe was ascertained, the salubrity of the air

* Dalton has given 31 grs. as a nearer approximation, and Prout states it at 31.0117 grs. These numbers coincide so nearly, that 31 will probably soon be adopted by all chemists, if confirmed by succeeding experiments.

in any situation was considered proportional to the quantity of oxygen present; the instrument used for ascertaining this was termed a EUDIOMETER. Instruments of this kind are still employed for determining the quantity of oxygen in various mixtures.

504. Berthollet considered that the elements of the air are retained together by chemical attraction; the proportions being in unison with equivalent numbers accords with this view.

505. Dalton maintained the opinion that they are merely mechanically mingled, proved that gases mingle mechanically that have no attraction, and shewed that even carbonic acid rises through a small tube into a bottle of hydrogen placed above it, though 22 times heavier than hydrogen, a corresponding quantity of hydrogen descending into the carbonic acid bottle. This proves a power of DIFFUSION among the gases previously unknown. Dalton concluded also that particles of the same gases repel each other, but that particles of different gases do not repel one another, and that one gas acts as a vacuum to another, though they diffuse themselves more slowly through each other than in a vacuum.

506. Mr Graham, in prosecuting this subject, has ascertained that each gas has a diffusive power or DIFFUSIVENESS peculiar to itself, which is inversely proportional to the square root of its density, and has drawn up tables representing their diffusive power, air being taken as a standard of comparison. In making his experiments, he generally used a tube with the gas under examination, open at one end, and closed with plaster-of-paris at the other, the diffusion taking place readily through the pores of this substance when moderately dry; it also takes place through membranes, small cracks in glass-vessels, and through numerous porous bodies. From all these considerations, Dalton's view of the constitution of the air is now generally adopted.

507. The CHEMICAL ACTION of air depends principally on the oxygen it contains. Deprived of this it can neither support combustion nor respiration. The manner in which the nitrogen acts, besides diluting the oxygen, has not been so distinctly traced. The influence of the moisture and carbonic acid is very important in numerous operations.

508. The oxygen of the air is largely consumed in many chemical operations; it is believed to be replaced principally by the action of the leaves of plants on carbonic acid, which in the sunshine absorb that gas and restore oxygen.

Combustion of Hydrogen with Oxygen and with Compounds of Oxygen and Nitrogen.

a.	H&O = ·H	1 eq. hydrogen & 1 eq. oxygen = 1 eq. water.
b.	H&·N^2 = ·H&N^2.	1 eq. hyd. & 1 eq. air = 1 eq. water + 2 eqs. nitrogen.
c.	H&·N = ·H&N.	1 eq. hyd. & 1 eq. nitrous ox. = 1 eq. water + 1 eq. N.
d.	2H&:N = 2·H&N.	2 eq. hyd. & 1 eq. nitric ox. = 2 eq. water + 1 eq. N.

CHAP. IV. SULPHUR.

509. Symb. S.—Eq. 16.—Sp. gr. 1.99. Melts at 216°, very fluid at 250°, viscid at 450°, rendered a little thinner by more heat, sublimes at 600°; crystallizes on cooling from the liquid state.

510. Solid, of a yellow colour passing into an amber when melted, brittle, a bad conductor of heat. In the form of minute crystalline grains, commonly called FLOWERS OF SULPHUR, when sublimed; WHITE when precipitated from solution in water (see 540), a small portion of this fluid being combined with it in this condition; it is then called LAC SULPHURIS. Insoluble in water, soluble in alco-

hol when they meet in vapour and are condensed, soluble also in oil of turpentine and other oils.

511. Poured into water when merely melted, it becomes solid, and presents no peculiar appearance; but if poured into water when at a higher temperature, it remains transparent and elastic for a considerable time after it cools, regaining eventually its ordinary form.

512. Sulphur takes fire at 300°, burning readily in air or oxygen, and forming sulphurous acid. Combines readily with numerous inflammables and metals, forming SULPHURETS; heat and light frequently appear as it combines with metals.

513. Procured principally from volcanic countries, and also by heating iron-pyrites (bisulphuret of iron).

514. Occasionally it is mixed with earthy matters; heat expels all the sulphur, and they are left.

SECT. I. *Compounds of Sulphur and Oxygen.*

515. HYPOSULPHUROUS ACID.—Symb. $:S^2$ = sulphur 32 + oxygen 16?—Eq. 48. Formed when a sulphite (a compound of sulphurous acid) is boiled with sulphur, an additional quantity of sulphur combining with it. Produced also when sulphurous acid dissolved in water is agitated with metallic iron, the metal being oxidated by the acid, and hyposulphite of iron formed in solution. It is obtained with difficulty from its combinations, being generally resolved, according to Sir J. Herschel, into sulphur and sulphurous acid, on adding a stronger acid to detach it. Particularly distinguished by its salts in solution dissolving newly precipitated chloride of silver, and acquiring an intensely sweet taste. Salts of this acid have long been termed Sulphureted Sulphites, by Vauquelin, Gay Lussac, and Dalton.

516. Dr Thomson describes a compound of sulphur 16 + oxygen 8, eq. 24, which he has termed SULPHUROUS ACID; and his hyposulphurous acid is regarded as a compound of 32 sulphur + oxygen 8.

517. SULPHUROUS ACID.—Symb. $:S$ = sulphur 16 + oxygen 16.—Eq. 32 by W, □ by V,—Sp. gr. 2.222.—W. of 100 c. i. 67.776 grs. Condensed into a liquid by a pressure of 2 atmospheres, or simply by cooling in a freezing mixture of salt and snow; the liquid acid boils rapidly at 14°, producing great cold, and is a good conductor of electricity, according to Mr Kemp.

518. Gaseous, transparent, colourless, odour pungent and suffocating; water absorbs 33 times its volume, the mercurial trough therefore must be used in collecting it; the solution has the same odour as the gas. It is not inflammable, nor does it support combustion. Decolorizes many vegetable and animal colouring matters, reddening some of the blues previously: the tints may often be recalled by a stronger acid, or by some salifiable bases.

519. PREPARED by burning sulphur in air or oxygen, $S\&O^2 = :S$; by heating sulphur with peroxide of manganese, when part of the sulphur combines with a portion of the oxygen; or by decomposing sulphuric acid by mercury or charcoal. Both charcoal and mercury take oxygen readily from sulphuric acid on applying heat; the charcoal taking oxygen from 2 eqs. of acid forms carbonic acid, which mixes with the sulphurous acid. The mercury decomposes 1 portion of sulphuric acid, combining with part of its oxygen, and separating sulphurous acid, while the oxidated metal unites with a portion of undecomposed sulphuric acid. (See Bipersulphate of mercury.) Mercury is always preferred where pure sulphurous acid is required.

520. SULPHURIC ACID.—Symb. ⁝S = sulphur 16 + oxygen 24.—Eq. 40.—Melts at 68°, boils at 120°. Solid, tough, produces white fumes when in contact with air, combining with its moisture; unites with water in all proportions. Called frequently ANHYDROUS SULPHURIC ACID.

521. Prepared by combining aqueous sulphuric acid in excess with oxide of antimony, expelling the water and part of the acid by a moderate heat, and then heating the residue to dull redness, collecting the distilled anhydrous acid in a cool receiver. Prepared also from dried sulphate of iron by heat: the product obtained at first consists of sulphuric acid, sulphurous acid and water; by heating this gently, the anhydrous sulphuric acid may be distilled into a receiver; the sulphurous acid arises from part of the sulphuric acid being decomposed by the oxide of iron, which becomes peroxide of iron.

522. AQUEOUS SULPHURIC ACID.—Symb. ·H⁝S = water 9 + sulphuric acid 40.—Eq. 49.—Sp. gr. 1.8485.—Boils at 590°; resolved by a red heat into sulphurous acid and oxygen.—Synon. OIL OF VITRIOL. Liquid, of an oily appearance, colourless, inodorous, intensely acid, corrodes animal and vegetable substances, produces great heat on combining with water; 1 acid + 4 water rise in temp. to 300° when mixed; condensation accompanies the combination. Freezes at —15°. Diluted acid, sp. gr. 1.78, congeals readily at 32°, and remains solid even at 44° or 45°. Diluted to a greater degree, a more intense cold is required to freeze it. Exposed to the air, it absorbs watery vapour, and becomes much diluted. Reddens powerfully vegetable blues.

523. Sulphuric acid has a strong affinity for salifiable bases, from which it can displace most other acids. Few agents are so extensively employed. The following Table shews the quantities of liquid acid, sp. gr. 1.8485, and of dry acid, in 100 parts of dilute acid of various specific gravities:—

Specific Gravity.	Liquid Acid.	Dry Acid.	Specific Gravity.	Liquid Acid.	Dry Acid.
1.8485	100	81.54	1.3884	50	40.77
1.8115	90	73.39	1.2999	40	32.61
1.7120	80	65.23	1.2184	30	24.46
1.5715	70	57.08	1.1410	20	16.31
1.4860	60	48.92	1.0682	10	8.15

524. Prepared commonly by the oxygenation of sulphurous acid. This gas, introduced into a large leaden chamber with moisture and nitrous or hyponitrous acid, forms a crystalline compound, which is resolved into sulphuric acid, nitrous acid, and binoxide of nitrogen, by a large quantity of water; the binoxide, attracting more oxygen from the air, produces nitrous acid, which contributes, as at first, to the production of more sulphuric acid. The weak sulphuric acid is concentrated by boiling, the first portions of water being expelled in leaden boilers, and the last in vessels of platinum or glass. Hyponitrous acid can oxygenate 1 eq. of sulphurous acid.

Fig. 7

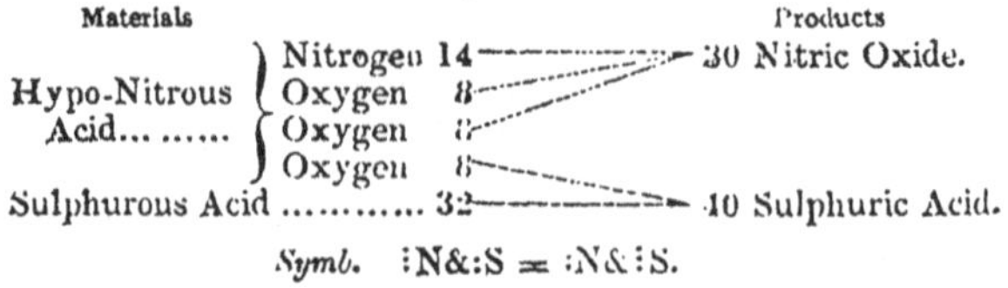

Symb. ⁝N&:S = :N&⁝S.

Nitrous and sulphurous acids were supposed to produce nitric, oxide, and sulphuric acid.

Fig. 8.

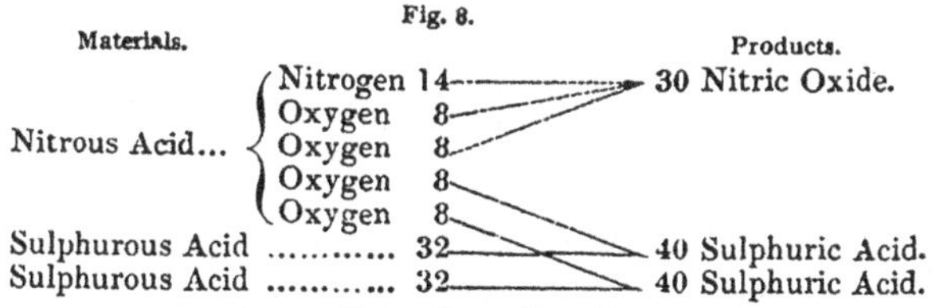

Symb. ·⁝N&2:S = :N&2⁝S.

But De Cleuchry considers that hyponitrous and sulphuric acids are formed first, the hyponitrous acid being resolved, when detached, into nitric oxide, and nitrous acid.

525. From the mode in which it is prepared, it often contains minute portions of sulphate of lead and sulphate of potassa; if required perfectly free from these, it must be distilled; when diluted with water, the sulphate of lead is deposited, rendering the liquid turbid. By evaporating a portion to dryness in a platinum capsule, the amount of these impurities may be ascertained.

526. Tests.—Baryta in solution is preferred, sulphate of baryta being precipitated, which is white, insoluble in water, and in nitric and muriatic acids. If the precipitate be heated with charcoal at the blowpipe flame, the sulphuric acid is decomposed, and on moistening it with water, and touching it with a solution of a salt of lead, the sulphur renders the lead black. See the decomposition of sulphates by charcoal, in the preliminary remarks on metals, for an explanation of this process. Sulphuric acid gives a copious white precipitate with soluble salts of lead.

527. Hyposulphuric acid.—Symb. :⁝S^2 = sulphur 32 + oxygen 40.—Eq. 72. It is always combined with water, and is resolved into 32 sulphurous acid and 40 sulphuric acid, when it is attempted to remove the water.

528. Formed when sulphurous acid gas is passed into water containing peroxide of manganese in suspension; the hyposulphuric acid combines with the oxide left, part of the oxygen of the peroxide being consumed in producing it. By baryta, the oxide of manganese is separated, hyposulphate of baryta being formed in solution, and from this the baryta is separated by sulphuric acid, as sulphate of baryta.

Preparation of Hyposulphate of Manganese.—Sulphurous acid 3 eqs. & peroxide of manganese 2 eqs. = hyposulphate of manganese and sulphate of manganese. *Symb.* 3:S&2:Mn = :⁝S^2·Mn&⁝S·Mn.

Sect. II. *Compounds of Sulphur and Hydrogen.*

529. Sulphureted Hydrogen or Hydrosulphuric Acid.—Symb. HS = hydrogen 1 + sulph. 16.—Eq. 17 by W, □ by V.—Sp. gr. 1.180.—W. of 100 c. i. 36.00 grs.

530. Gaseous, transparent, colourless; odour strong and offensive, like that from eggs in a state of putrefaction; does not support respiration nor combustion. Inflammable; 17 HS consume 24 oxygen, and the products are 32 sulphurous acid + 9 water. With air, a feeble and lambent blue flame is observed when this gas burns, part of the sulphur being generally deposited unconsumed. Mixed with oxygen in the above proportion, it detonates, and is all consumed. Noxious,

and even poisonous when inhaled, though mixed with much air. Air containing 1-150th of this gas kills a horse.

Fig. 9.

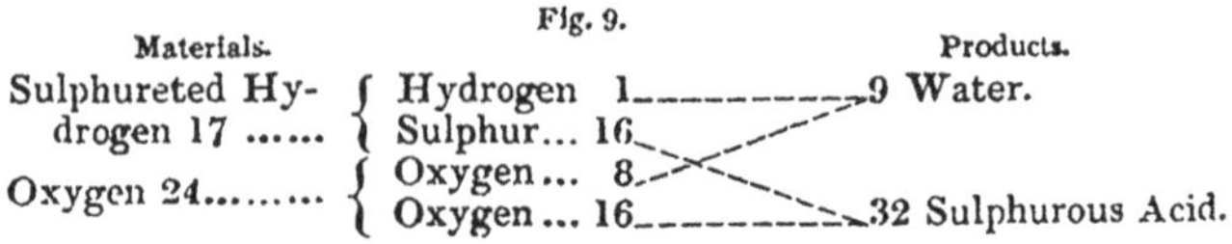

Symb. $HS\&O^3$ = ·H&:S.

531. Water 100 c. i. can absorb 253 c. i. according to Saussure. The solution acquires its taste and smell, reddens litmus, as acids; the gas is expelled by boiling or exposure to the air. A solution for ordinary purposes is easily procured by filling a bottle with water, introducing gas till it is 1-3d full, and then agitating it briskly. It must be carefully stopped, otherwise much gas will escape.

532. The oxygen of the air slowly decomposes this gas, combining with the hydrogen, and causing a deposition of sulphur. Some compounds of oxygen, as fuming nitrous acid, decompose it rapidly in the gaseous state with combustion. Chlorine, iodine, and bromine, combine with its hydrogen, and precipitate the sulphur. Three measures of sulphureted hydrogen, mixed with two of sulphurous acid over mercury, condense into a solid; the *hydrosulphurous acid* of Dr Thomson.

533. When these gases meet in a nascent state, they frequently decompose each other, the hydrogen of the one combining with the oxygen of the other, while the sulphur of both is precipitated.

534. Sulphureted hydrogen acts powerfully on many metallic oxides and solutions of the metals, a metallic sulphuret and water being formed; from the colour of the sulphurets being often very marked, it is much employed in detecting metals. Carbonate of lead is considered the most delicate test of this gas, being rendered brown by air containing 1-20,000dth part; when the quantity is larger, a black tint is produced. The carbonic acid, one of the ingredients of the carbonate, is disengaged in the gaseous form.

535. Prepared from sulphuret of iron 44, + water 9 + sulphuric acid 40 (aqueous acid 49 is used), with an additional quantity of water, 4 or 5 times the bulk of the sulphuric acid used, to prevent the process from being stopped by the sulphuret becoming covered with a crust of sulphate of iron. The products are sulphureted hydrogen and oxide of iron, which, with the sulphuric acid, forms sulphate of iron.

Fig. 10.

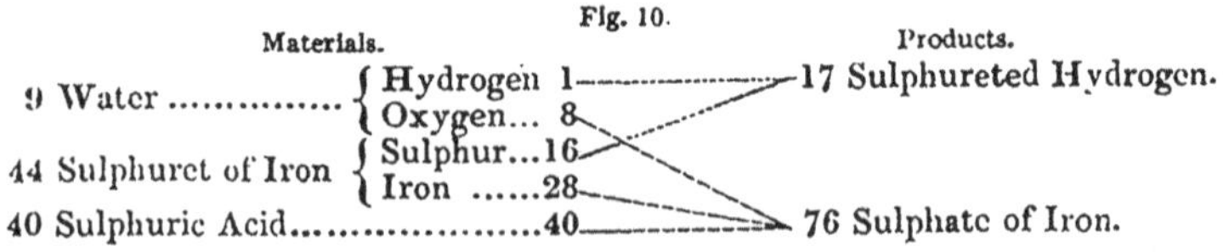

Symb. ·H&SFe&⁝S = ⁝S·Fe&HS.

536. Sulphuret of antimony and sulphuret of calcium are frequently substituted for sulphuret of iron, muriatic acid being then used instead of sulphuric.

537. BISULPHURETED or PERSULPHURETED HYDROGEN. Symb. HS^2 = hydrogen 1 + sulphur 32. Eq. 33.—From Thenard's researches, it is probable that there are many compounds of sulphur and hydrogen besides this and sulphureted hydrogen.

538. Liquid yellow, viscid or oily. Odour pungent, offensive, sulphureous

Soon decomposed by heat, or by exposure to the air; sulphureted hydrogen escapes, and sulphur is left. Charcoal, ammonia, many metals and metallic oxides, decompose it with rapidity; oxides of gold and silver become incandescent, and are reduced. It thus resembles much binoxide of hydrogen in facility of decomposition, and the phenomena which accompany it.

539. Prepared from the amber-coloured fluids formed by boiling potassa, soda, or alkaline earths, with water. Different compounds are produced in solution by the oxygen and hydrogen of the water uniting with different portions of sulphur.

Fig. 11.

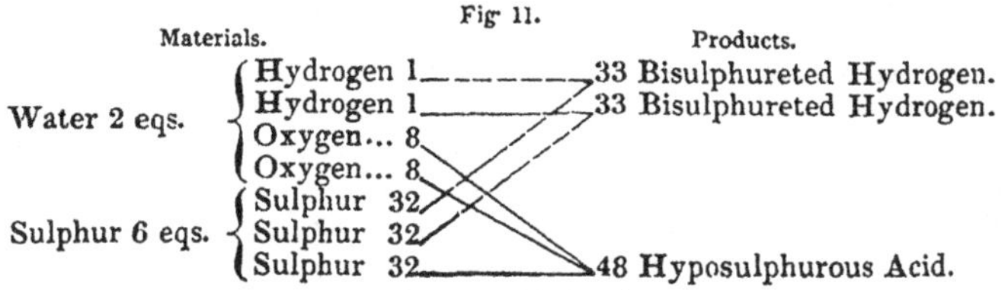

Symb. 6S&2·H = $2HS^2$&:S^2.

When the liquid is *poured into muriatic acid*, diluted with twice its weight of water, the salifiable base employed to promote the action of the sulphur on the water combines with the muriatic acid, and the bisulphureted hydrogen is slowly deposited. An excess of acid should always be present, otherwise it is very apt to be decomposed. Hence, also, the reason why the amber fluid must be poured into the acid, and not the acid into it.

540. The PRECIPITATED SULPHUR mentioned in the beginning of this chapter, is prepared from the amber-coloured liquid procured by boiling lime, sulphur, and water; muriatic acid is added as long as there is any precipitate. When the solution is diluted, and the acid added to it, no bisulphureted hydrogen falls along with the sulphur. A little sulphureted hydrogen is at the same time either expelled with effervescence or retained in solution.

Fig. 12.

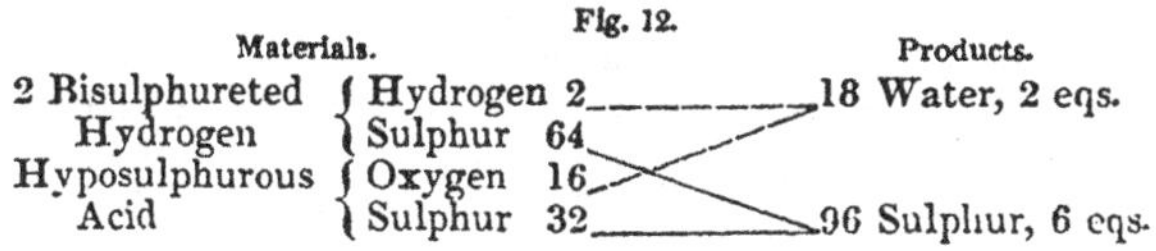

Symb. $2HS^2$&:S^2 = 2·H&6S.

CHAP. V. SELENIUM, BORON.

541. SELENIUM.—Symb. Se.—Eq. 40.—Sp. gr. 4·3.—Melts at 220°.—Boils at 600°.

542. Solid, dark-brown colour, semi-metallic lustre, imperfect conductor of heat and electricity, insoluble in water, inodorous, oxidates when exposed to the air and heated, producing a strong smell of horse-radish.

543. Selenium was discovered a few years ago, by Berzelius, in sulphur prepared from a compound of iron and sulphur, with which it was associated. It has not hitherto been met with in large quantities.

544. Its most important compounds are those which it forms with oxygen and hydrogen, as selenious acid, selenic acid, seleniureted hydrogen. They are very similar in all their general chemical relations to the compounds which sulphur forms with the same substances.

545. BORON.—Symb. B.—Eq. 8.—Nearly twice as heavy as water.

546. Solid, tasteless, inodorous, of a dark olive colour, insoluble in water, not fused by intense heat. Takes fire when heated to 600° in air.

547. Prepared from boracic acid by potassium, which removes the oxygen.

548. Boracic Acid.—Symb. :B = boron 8 + oxygen 16. Eq. 24.—Fused by heat, forming a glass as it cools.

549. Soluble in water; a hot saturated solution gives crystals of Aqueous Boracic Acid as it cools. Soluble in alcohol; the solution burns with a fine green flame.

550. It reddens feebly the vegetable blues, and turns turmeric brown, like an alkali. From its fusibility it is much used as a flux.

551. Prepared from the biborate of soda—the borax of commerce. By dissolving 152 parts in hot water, and adding 40 of sulphuric acid. Borax 152 = boracic acid 48 + soda 32 + water 72. The boracic acid separates in crystals as the liquid cools; 72 sulphate of soda remain in solution. The crystals must be washed with cold water, and, if required absolutely free from sulphuric acid, they must be heated in a platinum crucible till they are fused, and again dissolved and crystallized. Symb. 2:B+·Na&⁝S = ⁝S·Na&2:B.

CHAP. VI. PHOSPHORUS.

552. Symb. P.—Eq. 16.—Sp. gr. 1.7.—Melts at 100°.—Boils at 550°.

553. Solid, semi-transparent, colourless, or with a slight tinge of yellow; enters into slow combustion with the oxygen of the air at natural temperatures; gives a garlic odour, and becomes luminous in the dark, passing quickly into a state of rapid combustion if it be warm and in small chips; from the facility with which it is inflamed by friction, or a very gentle elevation of temperature, great care must be taken in operating with this element. It is best preserved under water, and in a dark place. Phosphorus burns brilliantly in air, and with great splendour in oxygen.

554. Phosphorus is dissolved in minute quantity by nitrogen gas, and this property enables it to act on the oxygen of the air, at a temperature below that (80°) at which it begins to affect pure oxygen gas. Mr Graham has found that air at 66° is prevented from acting on phosphorus by an admixture of 1-150th of sulphuric ether vapour, of 1-450th of hydruret of carbon, and of 1-4444th part of oil of turpentine vapour.

555. Preparation.—Bones are burned in an open fire to whiteness, reduced to powder, mixed with 3-4ths of their weight of sulphuric acid diluted with water, the mixture being afterwards heated gently for several days. The bones contain phosphoric acid and lime; the sulphuric acid unites with the greater part of the lime, and a solution of superphosphate of lime is separated by filtration; on evaporating it to dryness, mixing it with charcoal, and heating the mixture in a coated earthen retort, the charcoal takes away the oxygen from the phosphoric acid, forming carbonic oxide, which is disengaged; the phosphorus is converted into vapour, and condenses in water, into which it is led from the beak of the retort. It is purified by melting it, pressing it through shamoy leather under water, or by distillation.

Sect. I. *Compounds of Phosphorus and Oxygen.*

556. Oxide of Phosphorus?—When a stream of oxygen is directed upon phosphorus under boiling water, a portion of phosphoric acid is formed, which remains in solution, and an oxide of phosphorus, which has been regarded as a com-

pound of 3 atoms of phosphorus and 1 of oxygen. It is of a reddish-brown colour, and often appears in small quantity when phosphorus is inflamed in air.

557. Hypophosphorous Acid. — Symb. ·P^2 = phosphorus 32 + oxygen 8. Eq. 40.—Baryta, phosphorus, and water, being heated together, phosphoric and hypophosphorous acids are formed by part of the phosphorus taking oxygen from the water, and phosphureted hydrogen gas by another portion combining with its hydrogen. The gas escapes, and the acids combine with the baryta, the soluble hypophosphite of baryta being separated by filtration from the insoluble phosphate. Sulphuric acid precipitates the baryta, leaving the hypophosphorous acid combined with water. The solution is very acid, and becomes viscid and crystalizable by concentration. It is decomposed by heat; it is a powerful deoxidating agent, and all its salts are soluble.

558. Phosphorous Acid.—Symb. ⁝P^2 = phosphorus 32 + oxygen 24.—Eq. 56?

559. Prepared by heating bichloride of mercury with phosphorus, and decomposing by water the compound of chlorine and phosphorus distilled from this mixture. The hydrogen of part of the water combining with the chlorine, produces muriatic acid, which is expelled by heat, while the phosphorus, taking the oxygen, forms the phosphorous acid, which remains in combination with water. When evaporated to a syrupy consistence, crystals of the hydrated acid are procured.

560. Phosphorous acid mixed with phosphoric acid, is formed by the slow combustion of phosphorus at natural temperatures, or at a high temperature in rarified air.

561. Phosphorous acid is a powerful deoxidating agent like the hypophosphorous acid; heated in air, it burns and acquires more oxygen. The hydrated acid is decomposed by heat, forming phosphureted hydrogen and phosphoric acid.

562. Phosphoric Acid.—Symb. :⁝P^2 = phosphorus 32 + oxygen 40.—Some modifications of this acid have only of late years been distinguished by the researches of Dr Clarke and Mr Graham. Whenever the acid is strongly heated, it passes to the condition of Pyrophosphoric Acid, and still remains in this state for some days when dissolved by cold water, after which it returns to the condition of phosphoric acid. By boiling the solution this change is speedily effected. If the pyrophosphoric acid be combined with a salifiable base, it may be converted quickly into phosphoric acid by boiling in water to which some sulphuric acid has been added.

563. Mr Graham considers the modifications of phosphoric acid, which have been usually attributed to *isomeric* changes, to depend upon the acid combining with water in different definite proportions, and dissolving in an additional quantity of this fluid in these proportions. The following are the proportions of water associated with the common phosphoric acid, with the pyrophosphoric acid of Dr Clarke, and with the metaphosphoric acid which he has lately discovered.

	Phosphoric Acid.	Water.	Symbol.
Metaphosphoric acid, .	1 equivalent	1 equivalent	·H :⁝P^2
Pyrophosphoric acid, .	1 ...	2 ...	2·H + :⁝P^2
Common phosphoric acid,	1 ...	3 ...	3·H + :⁝P^2

564. Phosphoric acid is prepared by burning phosphorus in air or oxygen, by oxygenating phosphorus cautiously in nitric acid, and evaporating the solution in a platinum vessel to expel excess of acid, or by heating the phosphate of ammonia to redness, the ammonia being expelled. When prepared by combustion, it forms a light powder, falling in small flakes, and resembling snow. Procured by fusion, it resembles glass on cooling, and has been called *Glacial Phosphoric Acid.*

565. *Phosphoric Acid* is soluble, very acid, but not corrosive, volatilized by a strong heat, and solutions of its salts give a yellow precipitate with nitrate of silver, consisting of phosphate of silver. When converted into *pyrophosphoric acid*, the precipitate with nitrate of silver is white; *metaphosphoric acid* also gives a white precipitate with this substance, and it coagulates albumen, which is not done by the other forms of phosphoric acid.

SECT. II. *Compounds of Phosphorus and Hydrogen.*

566. PHOSPHURETED HYDROGEN OF HYDRURET OF PHOSPHORUS.—Symb. HP = hydrogen 1 + phosphorus 16; but, according to Rose, it consists of hydrogen 3 + phosphorus 32.

567. This gas is particularly distinguished by taking fire the moment it comes in contact with atmospheric air or oxygen; the products of the combustion are phosphoric acid and watery vapour, which ascend in beautiful wreaths when the gas is allowed to escape into the air through water. Another remarkable property of phosphureted hydrogen is its power of combining with various acids, and neutralizing them in the same manner as potassa, soda, or other salifiable bases. It is absorbed in large quantity by a solution of the common sulphate of copper or chloride of lime.

569. It is prepared by heating phosphorus along with water and potassa, soda, lime, baryta or strontia, being formed by the hydrogen of the decomposed water combining with part of the phosphorous; phosphoric and hypophosphorus acids are also produced, and combine at the same time with the alkali or earth employed.

569. Procured also by pouring muriatic acid diluted with water upon phosphuret of calcium; a portion of water is decomposed, and the calcium is oxidated by it as the gas is evolved, the hydrogen of the water and the phosphorus of the phosphuret combining together; the same action takes place without muriatic acid, but not so quickly; the muriatic acid unites with the lime or oxidated calcium as it is formed.

570. Another compound of phosphorus and hydrogen, called frequently BIHYDRURET OF PHOSPHORUS, on the supposition that it contains 2 eqs. of hydrogen + 1 of phosphorus, is said by Rose to be composed of the same proportions of elements as the preceding combination, and, therefore, *isomeric* with it. It is formed when aqueous phosphorous acid is decomposed by heat, the oxygen of one portion of the acid combining with the rest of the acid, and also with oxygen from the water, to form phosphoric acid; while its phosphorus, combining with the hydrogen of the water, produces the gas. It is not spontaneously inflammable, but burns brilliantly on the application of heat.

571. Mr Graham has lately shewn that the property of spontaneous inflammability may be communicated to the last species of gas by nitrous acid vapour, and that the first loses this property by the action of charcoal, phosphorus, and potasium.

SECT. III. *Compounds of Phosphorus and Sulphur.*

572. Phosphorus and sulphur, when heated, combine in any proportion. Small quantities, not exceeding a few grains, produce a violent and dangerous reaction during combination, unless the phosphorus be melted first in a tube, and the sulphur added in minute fragments at a time, taking care to boil it after each addition; the presence of a little water, from which the phosphorus and sulphur cannot easily be entirely freed, is the cause of the violence of the reaction, the phosphorus taking oxygen, and the sulphur forming with the hydrogen sulphureted hydrogen gas.

573. Compounds of these elements are usually termed PHOSPHURETS OF SULPHUR, or SULPHURETS OF PHOSPHORUS. They are more inflammable than phosphorus, and some of them take fire at natural temperatures when exposed to the air, more especially when they have been partially oxidated by allowing an imperfect combustion to go on for a short time.

CHAP. VII. CARBON.

574. Symb. C. Eq. 6. Carbon is not affected by heat when the air is excluded. It conducts heat imperfectly, but is a good conductor of electricity.

575. In its purest state it constitutes the DIAMOND; in the form it is usually met with it is called CHARCOAL, and is prepared from wood. IVORY BLACK is an impure charcoal prepared from bones, and often termed ANIMAL CHARCOAL. The cinder called COKE is a kind of charcoal prepared from coal. LAMP BLACK is the carbon deposited from oil, resin, and other compounds containing this element, when they are imperfectly consumed, causing the black smoke which then appears during their combustion. PLUMBAGO is composed of carbon with a very little iron.

576. Charcoal is solid, black, porous, and brittle. Inflammable, producing carbonic acid gas, when heated in air or oxygen. It often takes fire spontaneously when put into large heaps. It absorbs large quantities of a number of gases, amounting in some cases to nearly 100 times its bulk. The following table, from experiments by Saussure, shews the number of volumes of various gases absorbed in 24 hours by boxwood charcoal.

Ammonia, . .	90 vols.	Carbonic acid, .	35 vols.
Muriatic acid, .	89	Oxygen, . .	9.25
Sulphureted hydrogen,	65	Nitrogen, . .	7.5
Nitrous oxide, .	40	Hydrogen, . .	1.75

577. Charcoal is insoluble in water; destroys many animal and vegetable colouring matters, and the offensive taste and odour of putrescent vegetable and animal matter. It can also absorb or remove portions of numerous substances from solutions, as arsenious acid.

578. Charcoal is prepared from wood by heat, excluding the air. Wood is composed of carbon, oxygen, and hydrogen, and a small portion of earthy and saline matter. By the heat the oxygen and hydrogen are expelled, carrying along with them part of the carbon, and forming inflammable gases (composed of carbon with oxygen, and of carbon with hydrogen), water, acetic acid, tar, pyroxilic spirit, and kreosote. The excess of carbon forms the charcoal which remains. All animal and vegetable matters afford a portion of charcoal when they are subjected to a strong heat, air being excluded.

579. Charcoal, in its various forms, is used principally for fuel, producing a fixed red or white heat, without smoke or flame. Burned in an open fire, the intensity of the heat it produces is never great, being diminished by the influx of cold air on every side. Surrounded with fire-brick, or other bad conductors, the air being admitted solely from below by the furnace bars, the combustion is quicker, and the intensity of heat greater; if a high chimney be continued above such a fire-place, the column of heated air in the chimney being very light is unable to balance the cold air without, which presses quickly through the furnace bars, forcing up the air in the chimney, and being forced up in its turn in the same manner by the influx of more air. Hence, there is in this way a more rapid combustion, and a still greater intensity of heat produced, than in the pre-

ceding cases. By using bellows, fanners, or other apparatus to force in air, the heat may be still farther augmented; and, to produce the most intense heat, the air itself must be warmed in pipes by another furnace before it has access to the fuel. 6 parts of carbon require 72 of air (containing 16 of oxygen) for complete combustion, but the air does not act with the charcoal till heated to a certain temperature; therefore if not heated previously, its first effect upon the burning fuel is, to diminish its temperature by taking away the heat necessary to bring it into an active condition. By the great improvement of using HEATED AIR, introduced first in the Clyde Iron Works, an intense heat can be maintained with much greater facility than formerly.

580. In the furnaces in this country coke and coal are generally used instead of charcoal, except in peculiar operations.

581. Charcoal is used in a great variety of chemical operations besides those for the production of heat, and more particularly as a deoxidating agent.

SECT. I. *Compounds of Carbon and Oxygen.*

582. CARBONIC OXIDE, or OXIDE OF CARBON.—Symb. ·C = carbon 6 + oxygen 8.—Eq. 14 by W, □ by V.—Sp. gr. .9722.—W of 100 c. i. 29.652 grs.

583. Gaseous, transparent, colourless, inodorous, inflammable; 14 consume 8 of oxygen, forming 22 carbonic acid gas. Symb. ·C&O = :C. Burns with a pale blue flame. Does not support respiration or combustion. Water 100 c. i. absorb only 1.5. Does not affect the vegetable blues, nor lime-water; has no acid properties.

584. PREPARED from cr. oxalic acid by heating it with aqueous sulphuric acid. Cr. oxalic contains water, the sulphuric acid removes the water, and 36 oxalic acid are then resolved into 14 carbonic oxide + 22 carbonic acid. Both gases being collected together over lime-water, the lime absorbs the carbonic acid, and leaves the oxide.

Fig. 13.

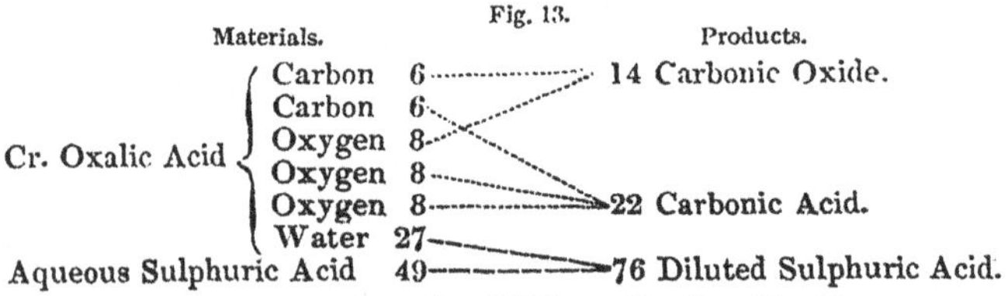

Symb. 3·H+⁝C²&·H⁝S = ·C&:C&4·H+⁝S.

585. When carbonate of lime, = carbonic acid and lime, is heated with iron-filings (or charcoal), part of the carbonic acid is decomposed by the iron as it is disengaged by the heat; lime-water removes the remaining carbonic acid.

Iron 28 & carbonate of lime 50 = oxide of iron 36, lime 28, & carbonic oxide 14.

Symb. Fe&:C·Ca = ·Fe&·Ca&·C.

Carbon 6 & carbonate of lime 50 = lime 28 & carbonic oxide 28, 2 eqs.

Symb. C&:C·Ca = ·Ca&2·C.

586. Carbonic acid gas passed repeatedly over charcoal, heated to redness in a tube, becomes carbonic oxide. Carbon 6 taking oxygen 8 from carbonic acid 22, forms carbonic oxide 14, while 14 more is left from the decomposed carbonic acid. The blue flame of a common cinder fire is produced by the combustion of the carbonic oxide formed in this manner: carbonic acid is formed at first, and this, rising

through the superincumbent red hot cinders, is decomposed by the carbon, and resolved into carbonic oxide, which becomes carbonic acid again as it burns.

Fig. 14.

Materials.			Products.
22 Carbonic Acid	Carbon	6	14 Carbonic Oxide.
	Oxygen	8	
	Oxygen	8	
6 Carbon........................		6	14 Carbonic Oxide.

Symb. C&:C = 2·C.

587. Carbonic Acid.—Symb. :C = carbon 6 + oxygen 16.—Eq. 22 by W, □ by V.—Sp. gr. 1.527.—W of 100 c. i. 46.596 grs.—Condensed into a liquid by a pressure of 36 atmospheres at 32°.—Synon. called Fixed Air by Dr Black, who discovered it.

588. Gaseous, transparent, colourless, cannot support respiration or combustion. When air is mixed, even with 1-5th of carbonic acid, a candle is extinguished in it; an oil lamp burns in air so loaded with it that a candle cannot.

589. Carbonic acid gas is absorbed by an equal bulk of water: common carbonic acid water (soda-water) is prepared with a much greater quantity of gas by subjecting it to pressure, having an alkali dissolved in the water. For experimental purposes, carbonic acid water is easily prepared by agitating equal bulks of carbonic acid and water in a bottle. It has an acidulous taste, reddens litmus, and precipitates lime water; it loses its carbonic acid when exposed to the air for some time, or boiled for a few minutes, and can, therefore, no longer redden litmus or precipitate lime-water.

590. In many fluids there is much carbonic acid retained by pressure or by a feeble attraction, and accordingly it escapes rapidly as they are poured from one vessel to another, or removed from bottles in which they have been tightly corked. The pungency of different fermented liquors depends upon the carbonic acid they contain, and the light and pleasant taste of spring water is owing to the same cause.

591. Carbonic acid is found extensively distributed throughout the globe. It exists in numerous minerals as a *Carbonate*, where it is united with salifiable basis, most frequently lime, magnesia, and oxide of iron. It occurs in a number of mineral waters in such quantities as to be the cause of all their peculiarities; in the air, it is met with at all heights and situations, excepting, perhaps, very near the surface of the ocean; and being a product of respiration, combustion, and other chemical processes, it often occurs in large quantity in particular situations.

592. The choke damp of miners, the heavy gas which gives rise to so many fatal accidents in breweries, the suffocating air that so frequently collects in old wells and pits, and which flows out in a stream from some caverns, as the Grotto del Cane in Italy, all consist of carbonic acid gas. Hence, to prevent accident, a lighted candle suspended by a string should be introduced into the air of any place suspected to contain carbonic acid gas, before any individual enters it; if the candle be extinguished, it would be dangerous and probably fatal to any one entering it. A certain proportion of carbonic acid mixed with air checks the combustion of a candle. Less, in general, is necessary than would prove injurious, at least immediately, to animal life. A larger quantity is required to extinguish a lamp. But its effects vary in different individuals; and accordingly, a candle burning in an atmosphere suspected to contain carbonic acid, cannot in all cases be regarded as a certain indication of its safety.

593. Prepared in a glass vessel, at the common pneumatic trough, by the action of muriatic acid diluted with water on carbonate of lime; chalk, marble, or

limestone, are generally employed, which consist principally, and sometimes entirely, of carbonic acid and lime. Carbonate of lime 50, contain carbonic acid 22 and lime 28. Muriatic acid 37 decompose 50 of the carbonate, the carbonic acid escaping as a gas, while the muriatic acid and the lime combine and produce 65 of the muriate of lime. *Symb.* :C·Ca&HCl = HCl·Ca&:C. Prepared also by the combustion of carbon in air or oxygen. Carbon 6 & oxygen 16 = carbonic acid 22. *Symb.* C&O^2 = :C.

594. Oxalic Acid, another compound of carbon and oxygen, will be described under Vegetable Substances.

Sect. II. *Compounds of Carbon and Hydrogen.*

595. Hydruret of Carbon.—Symb. HC = hydrogen 1 + carbon 6.—Eq. 7 by W, □ by V.—Sp. gr. .9722.—W of 100 c. i. 29.652 grs. Decomposed by a red heat, hydrogen and bihydruret of carbon being formed, while charcoal is deposited. Synon. Heavy Carbureted Hydrogen; olefiant gas, from its forming an oily compound with chlorine.

596. Gaseous, transparent, colourless, tasteless, inodorous, water 100 c. i. absorb 12.5. Inflammable, does not support combustion. Hydruret of carbon 7 consume 24 of oxygen as they burn, detonating violently if they be mixed previously in these proportions, the hydrogen 1 uniting with oxygen 8 and forming water 9, while the carbon 6 combines with oxygen 16 producing carbonic acid 22. It burns with a very luminous yellowish-white flame, the appearance varying much according to the manner in which it comes in contact with the air.

Fig. 15.

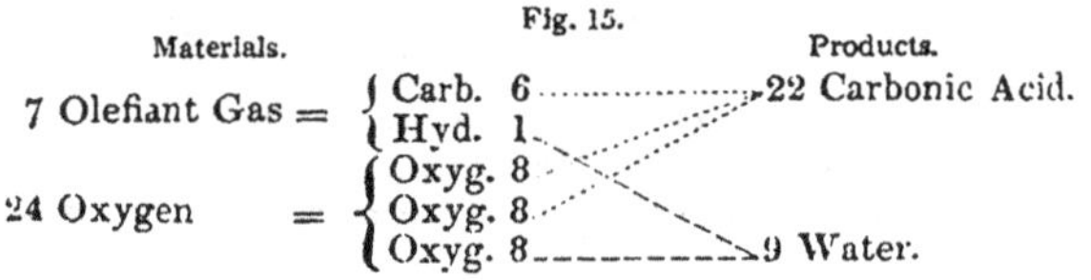

597. Prepared from coal, oil, resin, and substances of a similar nature, by the action of a red heat: as none of these, however, afford it at all pure, alcohol (spirit of wine), mixed with aqueous sulphuric acid, is preferred when it is required for experiment, though even then it is often associated with other gases. Alcohol 23 = oxygen 8 + carbon 12 + hydrogen 3; when heated in a retort with 3 times its bulk of aqueous sulphuric acid, then its oxygen 8 + hydrogen 1 form water 9, which combines with the acid, the remaining two of hydrogen uniting with the carbon, and producing 2 eqs. or 14 by weight of hydruret of carbon.

Decomposition of Alcohol so as to produce Hydruret of Carbon and Water, the Water combining with Sulphuric Acid not represented.

Fig. 16.

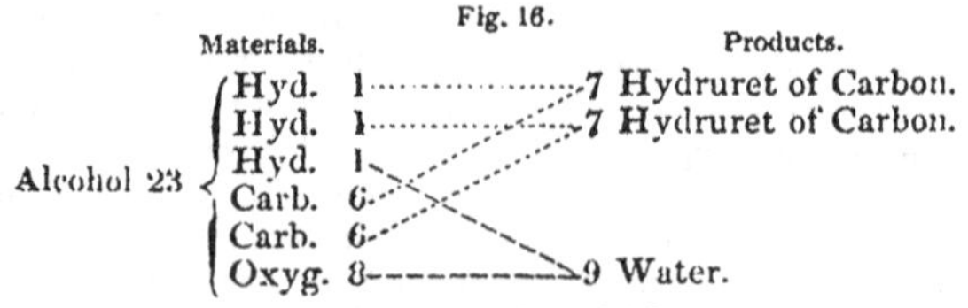

Symb. OC^2H^3 = ·H&2HC.

As the decomposition advances, the elements of the succeeding portions of alcohol arrange themselves in a different way, part of the carbon being then deposited, and bihydruret of carbon evolved, composed of the remaining carbon and hydrogen. But the reaction does not terminate here; the carbon soon acts upon a part of the sulphuric acid, taking away oxygen and disengaging sulphurous acid.

Fig. 17.

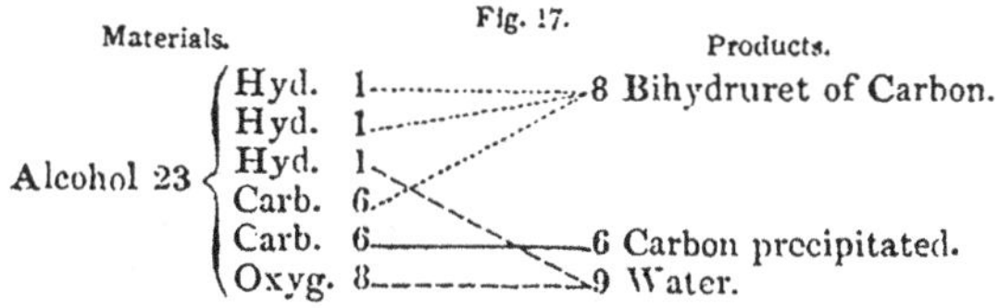

Symb. $OC^2H^3 = C\&\cdot H\&H^2C$.

598. Bihydruret of Carbon.—Symb. H^2C = hydrogen 2 + carbon 6.—Eq. 8 by W, □ by V.—Sp. gr. .5555.—W of 100 c. i. 16.944. Decomposed by a strong heat. Synon. Light Carbureted Hydrogen, Subcarbureted Hydrogen, the Fire Damp of Miners.

599. Gaseous, transparent, colourless, tasteless, inodorous; water absorbs 1-60th part of its bulk. Does not support respiration or combustion. Inflammable, burns with a yellow flame less luminous than that of the hydruret of carbon; 8 parts consume 32 of oxygen, the 2 of hydrogen taking 16 and forming 18 of water, while the 6 of carbon also take 16 and become 22 of carbonic acid.—Symb. $H^2C\&O^4 = 2\cdot H\&{:}C$. It detonates on applying a light, if the air or oxygen required for its combustion be previously mixed with it.

600. Formed in various proportions along with hydruret of carbon, when coal, oil, and resinous substances, are subjected to heat; and procured most easily by stirring the mud in stagnant pools, and collecting the gas evolved. Some carbonic acid is always associated with it when obtained in the latter way; agitation with lime-water removes it. It is also found in large quantity in many coal-mines.

601. Quadro-carbureted Hydrogen.—Symb. C^4H^4 = carbon 24 + hydrogen 4.—Eq. 28. When oil-gas is condensed at the portable gas-works, a liquid is obtained at the same time from which this compound may be distilled by the heat of the hand; it must be condensed in a receiver reduced to the temperature of zero by a freezing mixture. It boils at a temperature below the freezing point of water, and is very inflammable. Its vapour is twice as dense as the hydruret of carbon.

602. Bicarbureted Hydrogen.—Symb. C^2H.—Eq. 13 = carbon 12 + hydrogen 1.

603. Prepared from the same fluid that yields the preceding compound, condensing from the vapour that is produced when it boils at the temperature of 180°. Very inflammable, burns with much smoke. It occupies, when in the gaseous form, 1-9th of the space its elements do when uncombined.

604. Coal and Oil Gases are procured from coal and oil, by heating them to redness while the air has no access to them. In oil-gas, there is more hydruret of carbon than in coal-gas, and this is the cause of its burning with a more brilliant flame. The greater the density of the gas, it is the richer in carbon, and the more luminous. Besides the hydruret and bihydruret of carbon, various other compounds, as the oily matters alluded to in the preceding paragraphs, are believed to be present in small proportion, contributing to the brilliancy of the flame during combustion.

605. Coal-gas is not only less illuminating than oil-gas, but frequently contains a portion of sulphureted hydrogen, derived from the sulphur or bisulphuret of

iron in the coal from which it is prepared; by the action of lime and water it is entirely condensed. Oil consisting principally of carbon and hydrogen, the gaseous compounds it affords are comparatively pure; but coal containing also a very large quantity of oxygen and nitrogen, various other compounds besides the hydruret and bihydruret are procured at the same time, as water, ammonia, and carbonic acid, in addition to the tarry matter which is formed.

606. NAPHTHA is an oil procured by distillation from coal-tar; free from oxygen, and containing hydrogen and carbon alone, when pure. One measure of its vapour contains 6 equivalents of carbon and 6 of hydrogen condensed together. It is very inflammable, and is used for preserving potassium, and dissolving caoutchouc.

607. NAPHTHALINE is a white crystalline solid, which is formed when naphtha is subjected to a moderate heat and slowly distilled. It melts at 180°, boils at 410°, and forms with sulphuric acid the compound termed SULPHO-NAPHTHALIC ACID. Dr Thomson regards naphthaline as a compound of hydruret of carbon and bicarbureted hydrogen.

608. In the coal and oil gases prepared for ordinary combustion, small portions of several of these combinations are believed to be present, communicating their peculiar properties, and improving the brilliancy of the light.

609. The FLAME of different coal and oil gases is very variable. The most intense and perfect light is produced when the gaseous matter is completely consumed, while a separation of charcoal in the solid form is observed in the flame itself, which is rendered highly luminous or incandescent as it burns. This effect is produced most completely in the gas jet or burner, prepared so that two opposing currents as they strike upon each other expand into a flat sheet of flame, spreading in a direction exactly at right angles to the opposing currents in which they meet.

610. In the argand burner, when gas or oil is used, a similar effect is produced by allowing the air to enter freely within, as well as on the outside of the inflammable matter. The smoke produced during the imperfect combustion of oil or gas arises from part of the carbon separated not being consumed.

611. During the combustion of the tallow, wax, or oil, of a candle or lamp, gas is first produced by the decomposition of a portion upon the hot wick.

612. A most intense heat is produced by flame, and a very great heat is necessary for its production. A platinum wire or plate, if very thin, may be introduced when red hot into an explosive mixture without detonating it, maintaining however a slow combustion, by which sufficient heat is extracted to keep the platinum red hot.

613. The heat produced by gaseous bodies in a state of combustion, by which ame is formed, is very great. Iron-filings allowed to fall into the flame of a common lamp or candle immediately take fire.

614. Flame cannot pass through wire-gauze, the gaseous mixture composing it being so much cooled during its passage through the net-work, if the apertures be sufficiently small, that it ceases to be flame; nor can it set on fire a gaseous mixture on the other side of the gauze through which it may pass. It was on this principle that Sir H. Davy constructed his Safety Lamp, which consists of a common oil lamp having the wick covered with a wire cage; the oil burning within cannot set fire to the gaseous inflammable mixture in which the lamp is so often placed. It may enter the wire cage through the lamp, and burn there, but it cannot set fire to the gas without.

615. The flame of a common lamp or candle must be regarded as a mass of gaseous matter burning only on the outside, where it is in direct contact with the

air. If one end of an open glass tube be placed in the interior, and the other end elevated, inflammable gas will pass through it, which may be kindled at the open extremity of the tube without the flame.

616. EXPLOSIVE MIXTURES of inflammable gases and air are formed whenever the quantity of air mixed with the gas is sufficient for its combustion; the flame travels rapidly in it, the gas being supplied with all the air necessary.

617. With an excess of air, the action is more or less feeble; and if the quantity of air be large, the gas may burn where it is brought in contact with flame, but there alone, the particles of inflammable matter being too distant to inflame each other.

618. With an excess of the inflammable gas, the combustion is equally prevented from going on with explosive rapidity, as each portion not supplied with sufficient air must be mixed with a proper quantity before it can burn.

619. The FIRE DAMP of miners, which is disengaged from the coal in a number of mines, produces an infinite variety of explosive mixtures such as the above, as it escapes from the coal into the mine. By using the safety-lamp instead of a common candle, the miners can work in safety, though an explosive atmosphere may have been formed in the upper part of the mine, which would be instantly detonated by a common candle.

620. Sir H. Davy found that fire-damp does not explode unless mixed with more that 4 times its bulk of air; till 7 times its bulk have been added the explosion is feeble; with more than 14 there is no explosion at all.

621. The fire-damp is believed to have been formed in the coal by heat, and to be disengaged when an outlet is made for it by the miner striking it with his pick.

622. To VENTILATE MINES where noxious gases occur, two shafts or openings are generally made into each mine, one at each end of the workings; a fire placed at the bottom of one of these shafts causes an ascending current, which is supplied by air descending at the other shaft, traversing the whole of the workings, and maintaining the air sufficiently pure, all noxious gases being diluted and carried away in the current.

623. BLOWPIPES are small conical tubes bent at one end, so as to be easily introduced into the flame of a lamp or candle, directing it in a stream by blowing through them on any object to be heated. With this valuable instrument, numerous investigations are much facilitated, as most furnace operations can be imitated with it on a small scale. The greatest heat is at the point of the blue cone that is seen within the rest of the flame; the outer part of the flame beyond the blue flame is termed the OXIDATING FLAME, being mixed abundantly with the oxygen of the air; the inner portion is called the DEOXIDATING FLAME, being mixed with an excess of hot inflammable matter, which is a powerful agent in removing oxygen.

SECT. III. *Bisulphuret of Carbon.*

624. Symb. S^2C = sulphur 32 + carbon 6. Eq. 38.—Liquid, transparent, and colourless. Very volatile; odour fetid and offensive. Inflammable. Sulphurous and carbonic acids are the products of its combustion. Prepared by passing sulphur in vapour over charcoal heated to redness in a porcelain tube.

CHAP. VIII. CHLORINE.

625. Symb. Cl.—Eq. 36 by W.—□ by V.—Sp. gr. nearly 2.5.—W of 100 c. i. 76 grs. and a fraction.—Condensed into a liquid by the pressure of 4 atmospheres at 50°. Synon. Oxymuriatic acid, or oxygenated muriatic acid; dephlogisticated marine acid. The first two names were given when chlorine was regarded as a compound of muriatic acid and oxygen; an opinion which was defended with much ingenuity by the late Dr Murray. Sir H. Davy considered it an element, and gave it the name of Chlorine from its colour, derived from χλωϱος, green.

626. Gaseous, transparent, of a greenish-yellow colour; odour pungent, suffocating, extremely irritating and offensive. Uninflammable; supports combustion; destroys most animal and vegetable colouring matters and odorous effluvia, and decomposes a great number of vegetable and animal substances, from which it generally removes hydrogen, either directly, or by combining with the hydrogen of any water that may be present, and separating oxygen, which has the same effect.

627. Water absorbs $1\frac{1}{2}$ times its volume of chlorine, acquires its taste, colour, and smell, and also its bleaching properties. A crystalline compound is produced when it is introduced into ice-cold water, the HYDRATE OF CHLORINE, = water 90 + chlorine 36. Warm water, at temp. 90°, must be employed in collecting chlorine to prevent absorption; mercury combines with it, so that the mercurial trough cannot be used.

628. The solution of chlorine in water ought to be kept in dark bottles, and excluded from the light, otherwise the chlorine decomposes in a short time part of the water, combining with its hydrogen, and forming muriatic acid.

629. Numerous substances take fire when they are introduced into chlorine gas, as antimony and copper when in a minute state of division, turpentine and phosphorus. A candle with the flame blown out, but with the wick still red, is rekindled in chlorine, as Mr A. Trevelyan pointed out. In all cases where it supports the combustion of common inflammable matter, it combines generally with the hydrogen alone, separating in the form of a black powder the carbon, for which it has but a feeble attraction.

630. PREPARATION.—From the common muriatic acid by heating it with peroxide of manganese, diluting it with 1-6th of its bulk of water, that the action may not go on too quickly. Muriatic acid 37 decompose peroxide of manganese 44, disengaging 8 of oxygen, combining with the 36 of oxide left, and forming muriate of manganese. The oxygen detached decomposes 37 of muriatic acid, from which it takes 1 of hydrogen and disengages 36 of chlorine. Three parts of the common acid, and one of the oxide, are generally taken.

Chlorine from Muriatic Acid and Peroxide of Manganese.

Fig. 18.

Materials.			Products.
37 Muriatic Acid	Chlorine........	36	36 Chlorine.
	Hydrogen	1	9 Water.
37 Muriatic Acid		37	
44 Perox. Mang.	Oxygen	8	
	Protox. Mang.	36	73 Mur. of Mang.

Symb. 2HCl&:Mn = HCl+·Mn&·H&Cl.

631. Another process for preparing chlorine is explained by the following diagram:

Chlorine from Chloride of Sodium, Peroxide of Manganese, and Sulphuric Acid.

Fig. 19.

Materials.			Products.
60 Chloride of Sodium	Chlorine	36	36 Chlorine.
	Sodium	24	
44 Perox. Mang.	Oxygen	8	
	Protox. Mang.	36	
40 Sulphuric Acid		40	72 Sulph. of Soda.
40 Sulphuric Acid		40	76 Sulph. of Mang.

Symb. ClNa&:Mn&2⁝S = ⁝S·Na&⁝S·Mn&Cl.

632. Test.—Chlorine, in solution, is distinguished from all other substances by its power of discharging the tints of vegetable colouring matter, and by the copious curdy precipitate which it produces in solutions of the nitrate of silver, composed of chlorine and metallic silver, which becomes of a dark colour quickly when exposed to the direct rays of the sun, and more slowly by diffused day-light.

Compounds of Chlorine.

633. Peroxide of Chlorine.—Symb. ⁝Cl. = chlorine 36 + oxygen 32.—Eq. 68 by W, □□ by V.—Sp. gr. about 2.36.—W. of 100 c. i. nearly 72 grains. Explodes violently at 212°, and often during its preparation, so that great care is necessary in operating with it.

634. Gaseous, transparent, of a very deep greenish-yellow colour; odour aromatic; many substances inflame in it, and phosphorus takes fire when it comes in contact with it even under water. When decomposed by heat alone, 2 measures expand to 3, heat and light being evolved as they separate.

635. Prepared from chlorate of potassa by aqueous sulphuric acid. Here it is supposed that sulphuric acid 4 eqs. decompose 2 eqs. of chlorate of potassa, forming 2 eqs. of bisulphate of potassa, as they detach 2 eqs. chloric acid. These lose each 1 eq. of oxygen, and become peroxide of chlorine. The 2 eqs. of oxygen which they give out combine with 1 of chloric acid, still in combination with potassa, and now forming the perchlorate of potassa.

Fig. 20.

Materials.			Products.
2 Equivalents of Chloric Acid	Perox. Chl.	68	68 Peroxide of Chlorine.
	Perox. Chl.	68	68 Peroxide of Chlorine.
	Oxygen	8	
	Oxygen	8	
Chloric Acid		76	92 Perchloric Acid.

Symb. 3:⁝Cl = 2·⁝Cl&·⁝⁝Cl.

When the sulphuric acid and the potassa must be represented, the symbols may be arranged in the following manner:

3:⁝Cl·K&4⁝S = 2(2⁝S+·K)&·⁝⁝Cl·K.

636. The Euchlorine of Davy, formerly called Protoxide of Chlorine, is now regarded as a mixture of chlorine and peroxide of chlorine. It is formed by the action of muriatic acid on chlorate of potassa. If we suppose 1 eq. of muriatic acid to take potassa from 1 eq. of the chlorate, detaching chloric acid which is decomposed by another eq. of muriatic acid, the products are chlorine, peroxide of chlorine, water, and muriate of potassa. Symb. :⁝Cl·K&2HCl=HCl·K&Cl&·⁝Cl&·H. This gaseous mixture is transparent, of a deep greenish-yellow colour; odour like burned sugar; water absorbs 10 times its volume; detonates by heat alone, the

elements separating with a flash of light; phosphorus, oil of turpentine, and numerous other inflammables, take fire in it.

637. CHLORIC ACID.—Symb. :⁝Cl. = chlorine 36 + oxygen 40.—Eq. 76.—Formerly called Hyperoxymuriatic Acid.

638. Chloric acid has been procured hitherto only in combination with water. It reddens the vegetable blues; is decomposed by a number of substances which have a great affinity for oxygen; inflames paper, and other combustible matters, when its solution is concentrated; and, by heat, is resolved into chlorine and oxygen gases, and perchloric acid, which may be condensed as a colourless liquid in a receiver. The compounds of chloric acid giving off oxygen with facility by the application of heat, promote the rapid deflagration of inflammable matter.

639. PREPARED from a solution of chlorate of baryta by sulphuric acid, which separates the baryta in the form of the insoluble sulphate. Chlorate of baryta is formed by passing chlorine through a solution of baryta; 6 eqs. of chlorine decompose 5 eqs. of water, as in the following table; and 6 eqs. of acid are produced, 1 of chloric and 5 of muriatic, each eq. of acid combining with 1 eq. of baryta, in solution:

Fig. 21.

		Water.		Chlorine.		New Compounds.
5 Equivalents of water = 9 × 5 = 45.	Hydrogen	1	+	36	=	37 muriatic acid.
	Hydrogen	1	+	36	=	37 muriatic acid.
	Hydrogen	1	+	36	=	37 muriatic acid.
	Hydrogen	1	+	36	=	37 muriatic acid.
	Hydrogen	1	+	36	=	37 muriatic acid.
	Oxygen	40	+	36	=	76 chloric acid.

Symb. 6Cl&5·H = 5HCl&:⁝Cl.

The liquid containing these salts being boiled with phosphate of silver, it affects the muriate of baryta alone, forming insoluble phosphate of baryta and chloride of silver, nothing being left in solution except the chlorate of baryta, the compound required.

640. PERCHLORIC ACID.—Symb. ·:⁝Cl. = chlorine 36 + oxygen 56.—Eq. 92.—Procured in the manner described in 638; or by distilling the perchlorate of potassa with half its weight of sulphuric acid, diluted with 1–3d part of water; it condenses in combination with water in the receiver. Perchloric acid has been recommended as a test of potassa, forming with it a very sparingly soluble compound.

641. HYDROCHLORIC or MURIATIC ACID.—Symb. HCl. = chlorine 36 + hydrogen 1.—Eq. 37 by W, □□ by V.—Sp. gr. 1.28.—W. of 100 c. i. 39.1 grs.—Condensed into a liquid by a pressure of 40 atmospheres at 50°.

642. Gaseous, transparent, colourless, acrid and suffocating, uninflammable; does not support combustion. Condensed with explosive rapidity by water, which can take up 480 times its volume; heat is evolved during the combination.

643. Combined with water it forms common muriatic acid, and is more correctly termed *Aqueous Muriatic Acid.* Synon. Spirit of Salt, Marine Acid. This compound has a greater density than water, emits acrid vapours on exposure to the air; is transparent and colourless when pure, but has usually a straw colour, from the presence of small quantities of chlorine, nitrous acid, or peroxide of iron. Sp. gr. of acid usually manufactured 1.17, containing about 34 per cent. of real acid. The annexed table shews the specific gravity of the liquid of various degrees of strength, from a more extended table by Dr Ure:—

Specific gravity.	Dry acid in 100 parts.	Specific gravity.	Dry Acid in 100 parts.
1.2000	40.777	1.0000	20.388
1.1802	36.292	1.0879	17.941
1.1620	32.621	1.0617	12.641
1.1410	28.544	1.0417	8.563
1.1266	24.466	1.0200	4.078
1.1082	22.019	1.0020	0.403

644. The following table is taken from one of Dr Thomson's, and shews the specific gravity of the acid when united with water in different atomic proportions :—

Specific gravity.	Real acid.	Atoms of acid.	Atoms of water.	Specific gravity.	Real acid.	Atoms of acid.	Atoms of water.
1.203	40.659	1	6	1.1060	22.700	1	14
1.162	33.945	1	8	1.0960	20.442	1	16
1.139	29.134	1	10	1.0860	18.590	1	18
1.1197	25.517	1	12	1.0780	17.501	1	20

645. Nitrate of silver is the best test of muriatic acid, producing with its chlorine the insoluble chloride described in 632. The oxygen of the oxide unites with the hydrogen of the muriatic acid and forms water, which remains with the nitric acid of the nitrate.

646. Compounds of muriatic acid and metallic oxides produce water and metallic chlorides, when heated; and metallic chlorides, when they dissolve in water, are believed by many to decompose a portion and give muriates of the oxides.

647. Prepared from chloride of sodium (common salt) by the action of aqueous sulphuric acid. Chloride of sodium 60 = chlorine 36 + sodium 24. When it decomposes 9 water, the chlorine combines with the hydrogen of the water, and the sodium with the oxygen, the products being 37 muriatic acid gas and 32 soda. The sulphuric acid, which must be present to promote this decomposition, unites with the soda and forms sulphate of soda ; the muriatic acid is conveyed to a receiver, and condensed by water kept quite cold ; the sulphate remains in the apparatus in which the materials are heated. In general an excess of sulphuric acid is employed, so that a portion of bisulphate of soda is produced, and the aqueous sulphuric acid is diluted with 1-3d of its weight of water, and cooled before mixing with the salt, that the action may go on temperately.

Fig. 22.

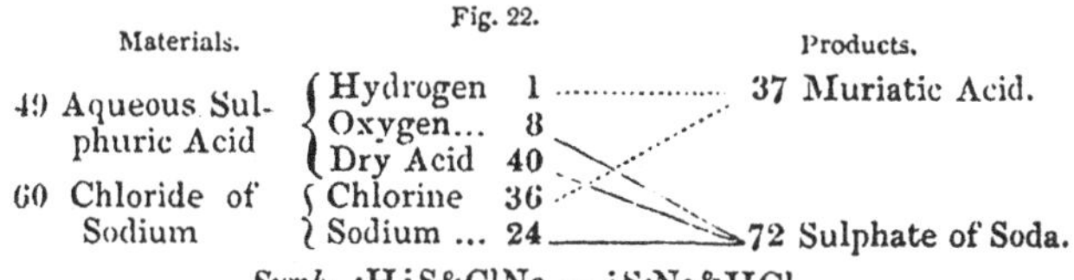

Symb. ·H⁝S&ClNa = ⁝S·Na&HCl.

648. Muriatic acid gas is easily procured from the liquid acid by heat, a large quantity escaping at a lower temperature than is necessary to boil water. It must be collected at the mercurial trough.

649. Muriatic acid gas may be formed on the small scale by applying a light to equal measures of chlorine and hydrogen : the mixture detonates, but there is no condensation unless water be present.

650. Nitromuriatic Acid, or Aqua Regia, is prepared by heating a mixture of nitric and muriatic acids. It contains a large quantity of chlorine, on which its power of dissolving gold or platinum depends, and for which principally it is used. The chlorine is derived from the muriatic acid, which is decomposed, its

hydrogen being removed by the oxygen of the nitric acid. Nitrous acid is at the same time produced, which gives it a very deep ruddy colour. It is generally prepared at the moment it is required, by mixing various proportions of the acids, according to the purpose to which it is to be applied.

Fig. 23.

Materials.			Products.
37 Muriatic Acid	Chlorine......	36	36 Chlorine.
	Hydrogen ...	1	9 Water.
54 Nitric Acid ...	Oxygen	8	
	Nitrous Acid	46	46 Nitrous Acid.

Symb. HCl&::N = ·:N&·H&Cl.

651. Quadrochloride of Nitrogen.—Symb. Cl^4N = chlorine 144 + nitrogen 14.—Eq. 158.—Sp. gr. 1.654.—Synon. Chloride of Nitrogen.

652. Liquid, of a yellow colour and oily appearance. Odour extremely penetrating, and almost insupportable; volatilizes rapidly at natural temperatures. Explodes with very great violence when heated to 200°, or on touching it with many oily substances, particularly olive oil and turpentine. Its elements are separated as it explodes, and the chlorine is believed to form muriatic acid gas with the hydrogen of a number of the inflammables which decompose it.

653. Chloronitrous Gas.—A compound of chlorine and binoxide of nitrogen prepared lately by Mr E. Davy, by the action of nitric acid on fused chloride of sodium.

654. Chloride of Sulphur.—Symb. ClS = chlorine 36 + sulphur 16.—Eq. 52.—It is formed by the action of chlorine on heated flowers of sulphur, and then distilled from any uncombined sulphur by a gentle heat.

655. Liquid, of a reddish colour, emits acrid fumes, boils at 280°; decomposes water, alcohol, and ether, the sulphur taking the oxygen, and the chlorine combining with hydrogen.

656. Another compound, with a larger quantity of chlorine, has been described.

657. Two compounds of chlorine and phosphorus have been described as the chloride and bichloride of phosphorus, but their precise composition appears still to be uncertain.

658. The Chloride or Protochloride of Phosphorus is formed when phosphorus is heated with the bichloride of mercury, and is separated by distillation. The Bichloride or Perchloride is obtained when phosphorus is introduced into chlorine gas, being formed by its spontaneous combustion.

659. Chloride of Carbon.—Symb. ClC = chlorine 36 + carbon 6.—Eq. 42. —Sp. gr. 1.552.—It is liquid, colourless, and boils at 170°. Discovered by Faraday, who prepared it by decomposing the succeeding compound by heat, a portion of the chlorine being expelled, and the chloride distilled.

660. Perchloride of Carbon.—Symb. Cl^3C^2 = chlorine 108 + carbon 12.— Solid, colourless, and transparent, fuses at 320°, and distils at 360°. Inflammable when heated; burns brilliantly in oxygen. Procured by Faraday from hydrocarburet of chlorine, all the hydrogen having been removed by repeated exposure to chlorine in the rays of the sun.

661. Bicarburet of Chlorine, called also Subchloride of Carbon, is a peculiar compound, which is procured when impure nitre and sulphate of iron are heated. The origin of the carbon has not been precisely pointed out.

662. Another chloride of carbon has been described by Liebig; it is formed when chloral is boiled in a solution of potassa, being converted into vapour and distilled. It is regarded as a compound of 4 eqs. of carbon and 5 of chlorine.

663. HYDROCARBURET OF CHLORINE.—Symb. H^2C^2Cl = hydrogen 2 + carbon 12 + chlorine 36.—Liquid, of an oily appearance; boils at 150°, completely decomposed by a red heat. Formed when chlorine and olefiant gas (hydruret of carbon) are mingled over water, condensing as an oily fluid which sinks in the water.

664. CHLORAL.—Discovered by Liebig, who regards it as a compound of 9 eqs. of carbon, 6 eqs. of chlorine, and 4 eqs. of oxygen. Sp. gr. 1.501. Boils at 200°.

665. Liquid, transparent, colourless, odour penetrating, almost tasteless, and has an oily appearance. It combines with water, sulphur, bromine, and iodine; and is decomposed when heated with different earths and metals, metallic chlorides being formed. It is prepared by passing chlorine into alcohol for a considerable time. The alcohol must be kept cool at first, and heated afterwards to expel the muriatic acid that is formed. A syrupy fluid is at length obtained, which becomes a white crystalline solid. Sulphuric acid removes the water it contains in this condition, when the chloral floats above it; by distilling it from lime it is separated from any adhering muriatic acid.

666. CHLOROCARBONIC ACID.—A compound of chlorine 36 + carbonic oxide 14. Discovered by Dr John Davy. It has also been called

667. PHOSGENE GAS, as it is produced solely by the action of light. In the sunshine the combination of the gases is rapid.

668. In concluding the history of chlorine, it is necessary to mention, that it was long regarded as a compound of muriatic acid and oxygen, an opinion ably defended by the late Dr Murray; the phenomena which it presents in its different combinations are all explicable on this supposition, though the view proposed by Sir H. Davy, and also by Gay-Lussac and Thenard, of its elementary nature, is considered more in accordance with actual experiment, and now almost universally adopted. In referring to chemical works published before the present view was generally entertained, the following memoranda will enable the student to translate the language formerly employed into the present nomenclature.

According to the old doctrine—	*According to Davy—*
1. Chlorine is a compound of muriatic acid 28 + oxygen 8 = 36.	1. Chlorine is an element.
2. Muriatic acid gas consists of 28 real acid + 9 water = 37.	2. Muriatic acid gas is the real acid, and contains no water, consisting of chlorine 36 + hydrogen 1 = 37.
3. Muriatic acid gas, acting on oxides, gives out its combined water, the real acid 28 combining with the oxide.	3. Muriatic acid gas acting on oxides is decomposed, its hydrogen uniting with the oxygen of the oxide, and producing the water which is detached, while a compound of chlorine and the metal is left.

CHAP. IX. IODINE.

669. Symb. I.—Eq. 124 by W, □ by V.—Sp. gr. of solid iodine 4.948; according to Dr Thomson 3.0844.—Sp. gr. of vapour 8.6.—W. of 100 c. i. 262.9 grs.—Name derived from ιώδης, purple or violet coloured.

670. Solid, opaque, of a bluish-black colour, and semimetallic lustre, odour pungent, taste acrid. Melts at 227°, and is vaporized at 350°, crystallizing as it cools.

It stains the skin of a yellowish-brown colour, destroys vegetable colours, and forms numerous compounds analogous to those of chlorine. At natural temperatures it evaporates slowly. Dissolves in 7000 of water, giving a deep brown colour to the liquid. Alcohol dissolves it more abundantly; the strong solution is almost opaque from the depth of colour.

671. Iodine is recognised with facility by the fine purple colour of its vapour, and the deep blue coloured precipitate with starch mixed with or dissolved in water. One part of iodine in 450,000 of water may be detected in this manner. Cold water must be employed in detecting iodine by starch, as the blue compound is decolorised by hot water.

672. Prepared from kelp, the ash that remains on incinerating sea-weed. When a solution of the soluble matter of kelp has been repeatedly crystallized, the drainings from each successive crystallization being boiled down to afford more crystals, they at length contain within a small bulk the iodine previously associated with a large quantity of kelp. The iodine exists there in the form of hydriodic acid (a compound of iodine and hydrogen), which is in combination with soda, and in solution. Sulphuric acid is added to detach the hydriodic acid, combining with the soda and forming sulphate of soda (which is separated by crystallization); sulphureted hydrogen, carbonic, and sulphurous acids, are expelled at the same time. Hydriodic acid is thus procured in an impure form, and in solution; by heating it with peroxide of manganese, one portion of the acid unites with protoxide of manganese, separating oxygen, which decomposes the rest of the hydriodic acid, by uniting with its hydrogen, and the iodine may be separated by sublimation. By using an excess of sulphuric acid, sulphate of manganese is formed, and more of the hydriodic acid decomposed.

Fig. 24.

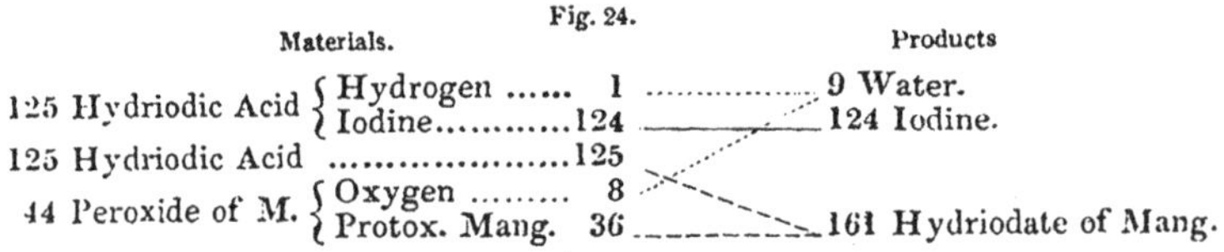

Symb. 2HI&:Mn = HI+·Mn&I&·H.

Compounds of Iodine.

673. Iodic Acid.—Symb. :⁝I = iodine 124 + oxygen 40.—Eq. 164.—Decomposed when heated to 300°.

674. Solid, semitransparent, soluble, deliquescent, acid and styptic taste, reddens and then destroys the vegetable colours. Forms detonating compounds with many inflammables, to which it communicates oxygen readily when heated. Its salts, which are called Iodates, are decomposed by heat; iodates of metallic oxides losing all their oxygen, and a compound of iodine and a metal being left.

675. Prepared by digesting iodine in nitric acid, as Mr Connell pointed out, the iodine being oxygenated by the decomposition of the nitric acid; or by passing protoxide of chlorine, prepared from muriatic acid and chlorate of potassa, over iodine in a tube, part of the iodine combining with the oxygen of the oxide and forming the acid, while the rest attaches itself to the chlorine and produces a compound, which is easily expelled by a moderate heat. When iodine is heated with water and an alkaline base, as potassa, soda, or lime, 6 eqs. decompose 5 eqs. of water, forming hydriodic and iodic acids, exactly in the same manner as in the reaction with chlorine, described in 639.

676. Oxide of Iodine, Iodous Acid, and Hyperiodic Acid, are other compounds of iodine and oxygen, of much less importance, which have been described lately.

677. Hydriodic Acid.—Symb. HI = iodine 124 + hydrogen 1.—Eq. 125 by W, □□ by V.—Sp. gr. 4.3.—W. of 100 c. i. 132.3 grs.

678. Gaseous, transparent, colourless, very acrid, produces white fumes when mixed with the air, must be collected over mercury, as water absorbs it with explosive rapidity. Its solution in water is easily concentrated by heat till it acquires a density of 1.5; it then boils at 260°. The solution is transparent and colourless at first, but acquires the colour of solutions of iodine by long keeping and exposure to the air and light, oxygen being absorbed and taking a portion of hydrogen, while the iodine previously united with it communicates colour to the solution. It reddens powerfully the vegetable blues.

679. Decomposed by chlorine, nitric and sulphuric acids, which liberate the iodine; the chlorine, and a portion of the oxygen of the acids uniting with the hydrogen; in the gaseous form, combustion ensues when nitrous acid and chlorine are used.—Chlorine and hydriodic acid = muriatic acid and iodine. Symb. Cl&HI = HCl&I.—Sulphuric acid and hydriodic acid = Sulphurous acid, water, and iodine. Symb. ⁝S&HI = :S&·H&I. Does not affect starch; mercury decomposes it, combining with the iodine and setting the hydrogen free.

680. Prepared in the gaseous form by the action of water on iodide of phosphorus. 50 parts of iodine may be mixed with 20 of water and 5 of phosphorus, and then heated; but with these materials, a slight explosion that ensues when the phosphorus and iodine begin to affect each other must be allowed to take place before heat is applied. The iodine taking hydrogen from the water produces hydriodic acid, and the oxygen unites with the phosphorus; when the process has continued for some time, phosphureted hydrogen is produced; it must then be stopped, as this gas is apt to explode in the retort, if the air mix with it.

681. A solution of hypophosphorous acid evaporated till it begins to be decomposed, and then mixed with an equal weight of iodine, gives very pure hydriodic acid, according to D'Arcet.

682. Sulphureted hydrogen gas, transmitted through water holding iodine suspended in it, produces quickly a solution of hydriodic acid gas, the iodine uniting with the hydrogen while the sulphur is deposited. Symb. I&HS = HI&S. Any excess of sulphureted hydrogen is expelled by heat, and the solution concentrated at the same time; the sulphur is removed by filtration.

683. Tests.—Starch indicates the iodine by the blue colour it produces after removing the hydrogen by sulphuric acid or chlorine water. Care must be taken to add the chlorine or sulphuric acid cautiously, as an excess prevents the blue compound of iodine and starch from being seen. Solutions of platinum produce a dark port-wine colour; salts of lead give the brilliant yellow iodide of lead; bichloride of mercury gives the bright red biniode of mercury. In all these cases, while the iodine is precipitated with the metal, the hydrogen of the hydriodic acid combines with the chlorine, oxygen, or other substance formerly united with it.

684. Teriodide of nitrogen.—Symb. I^3N = iodine 372 + nitrogen 14.—Eq. 386.—Called often Iodide of Nitrogen. Solid, black, detonates by heat or pressure. Prepared by the action of aqueous ammonia with iodine. Ammonia = hydrogen and nitrogen; one portion of iodine combines with the nitrogen, forming the teriodide; the other with the hydrogen produces hydriodic acid.

685. IODIDE OF SULPHUR is obtained by heating iodine and sulphur gently in a tube. It has a dark colour and radiated texture.

686. IODIDE OF PHOSPHORUS is procured by the action of iodine on phosphorus. They combine on being brought in contact, and so much heat is evolved that part of the phosphorus is inflamed if the air be not excluded. Iodine and phosphorus can combine in various proportions.

687. IODIDE OF CARBON is prepared from the periodide. Serullas procured the PERIODIDE by the action of iodine with alcohol.

686. CHLORIODIC ACID, or CHLORIDE OF IODINE, was procured by Davy on admitting chlorine into a receiver containing iodine, from which the air had been exhausted.

CHAP. X. BROMINE AND FLUORINE.

689. BROMINE.—Symb. Br.—Eq. 78?—Becomes solid at 4°, boils at 116°.—Sp. gr. 3.—Is a bad conductor of electricity. Its name is derived from βρωμος, signifying a strong and disagreeable odour.

670. The only element that is liquid at natural temperatures, with the exception of mercury. Of a deep hyacinthine red colour, volatilizes quickly, producing a ruddy coloured vapour, similar in appearance to nitrous acid. Has a strong taste and smell, and is very similar to chlorine in all its chemical relations. Corodes animal and vegetable substances, and is very poisonous. Bleaches vegetable colouring matter; supports the combustion of some inflammables. Soluble in ether, alcohol, and water; ether is preferred in extracting it by solution from various substances. Combined with water, it forms a hydrate, which crystallizes when cooled to 32°. With starch it forms a yellowish-coloured compound.

691. PREPARED from BITTERN, the drainings from common salt procured by boiling down sea-water. It exists there, united with hydrogen, in the form of hydrobromic acid, which is combined with magnesia, and retained in solution by water along with a number of different salts containing sulphuric and muriatic acids, soda, and magnesia. Chlorine is passed into the bittern so long as it deepens in colour; muriatic acid is formed by the chlorine combining with the hydrogen of the hydrobromic acid, and unites with its magnesia; the disengaged bromine communicates a yellow tint to the bittern. Symb. HBr·Mg&Cl = HCl·Mg&Br. By boiling the bittern the bromine is expelled, and must be collected in a receiver kept very cold.

692. Sulphuric ether agitated with the bittern dissolves it readily, and produces a liquid with a fine hyacinthine colour. The bromine is sometimes separated in this manner from the bittern; when the solution is agitated with a solution of potassa, its colour disappears; the bromine decomposes a portion of water, in the same manner as chlorine (see 639), and produces bromic and hydrobromic acids. These acids combine with the potassa, bromate and hydrobromate of potassa are formed; the bromate forms cubical crystals on concentrating the solution, which are easily separated. The bromine is then detached from these by chlorine, and distilled into a receiver.

693. Numerous combinations of bromine have been pointed out, all analogous in properties to the corresponding compounds of chlorine. Bromic acid = bromine 78 + oxygen 40; eq. 118; symb. ::Br.—Hydrobromic acid = bromine 78 + hydrogen 1; eq. 79 by W, □ by V; symb. HB.

694. Bromine appears to have a weaker affinity for different bodies than chlorine, but a more energetic power of attraction than iodine. It is found very uni-

versally diffused in the waters of the ocean, and in other mineral waters, though in very small quantity. It is believed to be generally in the form of hydrobromic acid; in this condition, it gives a white precipitate with solutions of lead and silver, composed of bromide of lead, or bromide of silver, from which bromine may be expelled in vapour by chlorine.

695. FLUORINE.—Dr Faraday has announced that he has procured this element; but the details of its properties and the mode of preparation have not yet been published.

696. HYDROFLUORIC ACID, commonly called FLUORIC ACID.—Sp. gr. 1.06.—Transparent, colourless, emits copious and very offensive fumes on exposure to the air, combining with its moisture. Has a great attraction for water, unites with it in all proportions, a hissing noise being at the same time produced. Reddens powerfully the vegetable blues; dissolves silica, forming a gaseous compound (see fluosilicic acid), and cannot therefore be prepared in glass or other vessels containing this substance. Very corrosive of animal and vegetable matter; a single drop falling on the skin corrodes it immediately, and produces an ulcer, which is accompanied by very great irritation, and healed with difficulty.

697. Hydrofluoric acid acts very energetically with a number of the metals, hydrogen being disengaged, and a metallic fluoride formed. It is used frequently for etching on glass, which is covered in the first place with wax, this being removed wherever it is intended that the fluoric acid should act; it is then covered with hydrofluoric acid, previously diluted with several times its weight of water, or exposed to the vapour of the acid.

698. PREPARED from fluoride of calcium (Derbyshire or fluor-spar), a compound of fluorine and calcium, by the action of aqueous sulphuric acid. Fluoride of calcium = fluorine and calcium; the fluorine with the hydrogen of the water forms the hydrofluoric acid; the calcium with the oxygen of the water produces lime, which remains with the sulphuric acid as sulphate of lime.

Fig. 25.

	Materials.		Products.
Aqueous Sulphuric Acid	Hydrogen	1	19 Hydrofluoric Acid.
	Oxygen	8	
	Sulphuric Acid	40	
Fluoride of Calcium	Fluorine.........	18	
	Calcium	20	68 Sulphate of Lime.

Symb. ·H⁝S&FCa = ⁝S·Ca&HF.

699. FLUOBORIC ACID GAS.—Regarded as a compound of fluorine and boron, or hydrofluoric and boracic acids, when acted upon by water, the fluorine taking hydrogen, while the boron combines with oxygen. Gaseous, transparent, colourless, reddens litmus, has a great attraction for water, producing a dense white cloud as it comes in contact with air containing moisture, and being absorbed in large quantity when introduced into this fluid. It is used occasionally to detect moisture in air or gases, producing dense vapours when there is any present.

700. The solution is acrid and corrosive, fumes on exposure to the air, and contains, when the water is saturated, 700 times its volume of gas. Potassium inflames when heated in it, combining with the fluorine and detaching the boron.

Prepared by heating, in a leaden retort, a mixture of fused boracic acid and fluor spar; or by heating a concentrated solution of boracic and hydrofluoric acids.

CLASS II. METALS.

701. METALS are particularly distinguished by their opacity, lustre, and power of conducting heat and electricity. They are distributed in large quantity throughout the globe, but appear in very various forms, according to the substances with which they are associated. They are seldom met with in a pure form, but must be separated chemically from their ores. Oxygen, sulphur, chlorine, and sulphuric, nitric, carbonic, and phosphoric acids, are the substances usually found in combination with them. They are said to be NATIVE when they occur pure, and MINERALIZED when they are combined with other bodies.

702. A great number of metals present a crystalline texture, and when cooled slowly on passing from the liquid to the solid form, so that a part in the liquid state may be poured off from the first portion that consolidates, distinct crystals can often be obtained with great facility.

703. DENSITY. Most metals, and also the greater number of their compounds, are very heavy compared with water; some metals, however, as potassium, are lighter than water.

704. MANY metals are MALLEABLE (can be beat out into thin leaves), and DUCTILE (drawn into fine wires). Those that have a highly crystalline texture, as antimony, arsenic, and bismuth, have not these properties. Some metals, as iron and platinum, are malleable and ductile at a high temperature; in this state two pieces of the metal can be united by hammering, an operation termed WELDING.

705. The TENACITY of metals varies very much; their relative tenacity is ascertained, by finding the weight which a rod or wire of a given length and diameter can support. If the tenacity of lead be taken as a standard of comparison, and reckoned 1, then zinc is represented by 4, gold by 5½, silver by 6½, platinum by 10, copper by 11, and iron by 20, omitting small fractions. Guyton Morveau found that an iron wire, little more than 2-3ds of a line in diameter, could support a weight of 549¼ lbs.

706. The points at which metals melt are very various. Mercury is fluid at —39.5°; platinum yields to the oxyhydrogen blowpipe, but resists the heat of our strongest furnaces.

707. Metals lose all their lustre, and assume the appearance of earths, when they enter into combination with the greater number of the NON-METALLIC BODIES. Hydrogen and carbon form with them few important compounds compared with the other non-metallic substances; no compound of nitrogen with a metal is known.

708. ACIDS do not combine with metals, but unite with their compounds with oxygen or other substances.

709. METALLIC OXIDES are generally formed by the action of metals on air, water, nitric acid, sulphuric acid, nitrates, chlorates; pure oxygen is rarely used. CALCINATION was the term used formerly to express the oxidation of a metal effected by the action of the air, and the oxide formed was denominated a CALX, from its being earthy like lime. The term calcination is now used in a more general sense by many chemists, and is applied whenever any solid matter has been subjected to a strong heat.

710. METALLIC SULPHURETS may be formed—I. By heating sulphur with many of the metals, a great increase of heat generally attending the combination, and frequently an evolution of light.

711. II. By the action of sulphureted hydrogen on numerous binary compounds of the metals, the sulphur combining with the metal, while the hydrogen unites with the other ingredient. Thus sulphureted hydrogen with oxide of lead produces sulphuret of lead and water.

712. III. By the decomposition of metallic oxides heated with sulphur, one portion of the sulphur combining with the oxygen, and another with the metal of the oxide. See diagram par. 774, p. 87.

713. IV. By the decomposition of SULPHATES of METALLIC OXIDES heated with charcoal or hydrogen, the latter being passed over them in a red hot tube. Each eq. of a sulphate contains 4 eqs. of oxygen, 3 in the sulphuric acid, and 1 in the oxide. Carbon 4 eqs., remove all the oxygen, each eq. of carbon combining with 1 of oxygen and forming 1 of carbonic oxide. Sulphates of the oxides of the fixed alkalis and earths are decomposed in this manner, but the sulphates of the common metallic oxides lose all their acid when heated to whiteness or a less elevated temperature, either with or without carbon. See diagram par. 773, p. 87.

714. PHOSPHURETS and SELENIURETS of the metals may, in many cases, be procured by processes similar to those for the preparation of the sulphurets, but they are of much less importance.

715. CHLORIDES, IODIDES, BROMIDES, and FLUORIDES? may be prepared in some cases by the direct combination of their elements, the evolution of heat and light very frequently attending their union.

716. They may also be obtained by the action of hydrochloric (muriatic), hydriodic, hydrobromic, and hydrofluoric acids, upon metallic oxides, the hydrogen of the acid uniting with the oxygen of the oxide, and forming water.

717. When metallic chlorides, iodides, bromides, or fluorides, decompose water, the chlorine, iodine, bromine, or fluorine, again acquires hydrogen; hydrochloric hydriodic, hydrobromic, hydrofluoric acids are produced, and the metal combining with the oxygen forms a metallic oxide.

718. Compounds of metals and cyanogen act in a similar manner.

719. When these metallic chlorides, iodides, &c. dissolve in water, it is uncertain whether they merely combine directly with the water, or whether they react on one equivalent of water, as stated at 717, the compounds so produced uniting with the rest of the water not decomposed.

720. Compounds of CHLORIC, IODIC, and BROMIC ACIDS with METALLIC OXIDES, lose all their oxygen by exposure to heat, nothing being left but the metal in union with the chlorine, iodine, or bromine.

721. Metals combine with each other, forming compounds termed ALLOYS; these are termed AMALGAMS when they contain mercury.

722. When metals are procured from their ores, or other compounds made artificially, they are said to be REDUCED, and the process is called REDUCTION. The term REGULUS is often employed to signify a pure metal. When mercury is reduced, it is commonly said to be REVIVED.

723. REDUCTION OF METALS.—I. Some metals may be reduced by heat alone, as gold, silver, mercury, and platinum, when combined with oxygen or various other substances.

724. If the heat given be not sufficient to melt the metal as it is reduced, it will appear as an earthy powder, having no metallic lustre, as when the muriate of ammonia and platinum is heated to redness, nothing being left but a gray

coloured powder, which assumes the metallic lustre when welded or melted at an elevated temperature.

725. II. When the reduction cannot be effected by heat alone, some substance must be introduced which shall assist in removing those bodies with which the metals may be combined. Carbon in different forms is generally employed to remove oxygen. Different metals, or compounds containing them, are much employed to separate sulphur, chlorine, iodine, &c. Iron, tin, potassium, lime which contains the metal calcium, and carbonates of potassa and soda, containing the metals potassium and sodium, are most in use. It is the metal that is the active ingredient in reducing the metallic ore, the different substances associated with the metallic compounds employed to promote the reduction being usually disengaged, or entering into new combinations with some of the ingredients of the metallic ore.

726. III. Metals are occasionally employed to reduce other metals from their combinations when dissolved in water. Thus iron precipitates copper from the sulphate in solution, the iron assuming the place of the copper, and combining with the sulphuric acid and oxygen previously associated with it. Even where the metallic compound to be reduced is insoluble in water, a similar decomposition may often be effected by boiling it in this fluid with the decomposing metal. Thus, zinc boiled in water with the chloride of silver, combines with the chlorine, and is dissolved, the silver being separated from the insoluble chloride of silver in a minute state of division.

727. Alkalis and other metallic oxides separate metallic oxides from their salts; metals are generally required to separate metals in the metallic form.

728. IV. In a great number of cases, when the reduction of metals is effected by the action of some decomposing agent at an elevated temperature, it is necessary to add some substance which shall melt easily, and allow all the metallic particles, as they are evolved, to subside in it and collect together. Bodies of this description are of great importance in the reduction of many metals, and are termed FLUXES, as they melt or flow during the operation of reduction. A flux may act in different ways; 1. merely by melting and allowing the metallic particles to fall through it; 2. it may combine as it melts with ingredients mechanically mixed in a minute state of division with the ore, and forming a very fusible compound with them, allow the reduced metal to fall through it; 3. it may affect chemically some of the ingredients combined with the metal in the ore, forming with them the fluid matter through which the metal is deposited.

729. It is not necessary that the flux shall in all cases be easily fused by itself. It is often sufficient if it form a fusible compound with the ingredients which it may come in contact with during the operation of reduction. Thus, no substance is more extensively employed as a flux than lime, though one of the most infusible of the earths, as it fuses when combined with the various substances that act upon it in many processes of reduction.

730. The greater the number of earthy or saline metals mixed together, the more fusible is the flux which they form.

731. A very fusible flux for experiments on a small scale, is made by mixing 5 parts of carbonate of potassa with 4 of carbonate of soda.

732. V. Metals may be reduced by electricity, the galvanic battery being preferred in these cases.—See Galvanism.

733. When reduction is effected in this manner, the metal always appears at the negative pole.

734. VI. The combined influence of galvanism and chemical attraction may at times be brought into play with advantage, by placing along with the compound

subjected to the galvanic current some substance which has a great affinity for one of its elements.

735. Before metallic ores are reduced, they are frequently subjected to heat for a considerable time by mixing them with inflammable matter, which is kindled and slowly burned away; in such cases they are said to be ROASTED.

736. ARRANGEMENT OF METALS.—The metals are divided into various groups by different chemists; and a particular order is often observed in treating of those described in each successive group, according to their relation to heat, electricity or power of attraction for other bodies. In this work, they are divided into three Orders: *

737. I. KALIGENOUS METALS, from Kali, the old name for potassa, used in a general sense to signify an alkali, and γιννάω, I produce. These include the metals Potassium, Sodium, and Lithium, which, with oxygen, form the alkalis Potassa, Soda, and Lithia.

738. II. TERRIGENOUS METALS, from Terra, the Latin for the earth, and γιννάω. This order includes all those metals which, with oxygen, form the substances called EARTHS by chemists. They are usually subdivided into *Alkaline Terrigenous Metals, and Common Terrigenous Metals*, the earths produced by the former resembling the alkalis in many respects.

1. Alkaline Terrigenous Metals.—Calcium, Barium, Strontium, Magnesium.
2. Common Terrigenous Metals.—Aluminum, Silicum, Glucinum, Ittrium, Zirconium, Thorinum.

739. III. CALCIGENOUS METALS.—These include the common metals, all of which can form a calx (see 709), or earthy-looking substance when combined with oxygen. They may be subdivided into—

1. Common metals whose oxides are not reduced by heat alone, viz.

Iron.	Bismuth.	Arsenic.	Cerium.
Lead.	Manganese.	Vanadium.	Columbium.
Copper.	Chrome.	Cadmium.	Tellurium.
Zinc.	Cobalt.	Tungsten.	Titanium.
Antimony.	Nickel.	Molybdenum.	Uranium.
Tin.			

2. Common metals whose oxides are reduced by heat alone, viz.

Mercury.	Platinum.	Rhodium.
Silver.	Palladium.	Iridium.
Gold.	Osmium.	Pluranium.

740. In describing the individual metals contained in each order or subdivision, they are arranged in the manner in which it is considered the student will acquire with greatest facility a knowledge of their chemical relations.

ORDER I. KALIGENOUS METALS.

741. The Kaligenous Metals form alkalis with oxygen, which are very soluble, acrid, and corrosive; have a peculiar taste; turn the vegetable blues to a green, and turmeric to a brown; neutralize the acids, and form soaps with oils, &c. Their salts are more soluble in general than those of the earths or common metals.

* In designating the different orders, I have used the terms proposed by Dr Hope.

CHAP. I. POTASSIUM.

742. Symb. K—Eq. 40—Sp. gr. 0.865.—Solid, hard and brittle at 32°; soft and malleable at 50°; perfectly fluid at 150°; volatilized at a temperature below a red heat.

743. Lustre white and brilliant like mercury, oxidates rapidly in the air, inflames when heated in the air or oxygen, or when brought at natural temperatures in contact with water, ice, liquid acids, and numerous other substances containing oxygen. Has a powerful attraction for non-metallic bodies, and can separate a number of other metals from their combinations with oxygen. From its power of attracting oxygen, it must be kept carefully excluded from air and moisture. It is preserved in naphtha which contains no oxygen; care must be taken in the selection of the naphtha, as some kinds either contain some oxygen as a constituent element, or have a portion of water adhering to them, acquired during their preparation, so that they oxidate slowly any potassium in contact with them.

744. Potassium, in acting on ice or water, produces potassa and potassiureted hydrogen. The potassa is formed by part of the potassium taking oxygen, and the hydrogen of the water combines with another portion, producing POTASSIURETED HYDROGEN, a gas which is spontaneously inflammable, taking fire in contact with the air; as it burns, the products are potassa and water, formed by the oxygen of the air uniting with its potassium and hydrogen. Another compound of potassium and hydrogen has been described by Thenard; it is solid, not spontaneously inflammable, and procured by heating potassium in hydrogen.

745. If potassium be plunged under water, potassa alone is formed, and hydrogen gas escapes.

746. PREPARATION.—Discovered by Davy, who obtained it by subjecting hydrate of potassa to the galvanic current, after moistening it slightly that it might be a better conductor. Potassa contains potassium and oxygen; the potassium appeared in minute quantities at the negative pole.

747. Gay-Lussac and Thenard prepared it more abundantly by melting hydrate of potassa, and allowing it to flow over iron-turnings at a white heat. Iron 28 & potassa 48 = oxide of iron 36 & potassium 40. *Symb.* Fe&$\dot{K}$ = $\dot{Fe}$&K. The iron decomposes both the potassa and the water, the oxygen of both being retained, while the potassium is volatilized, part of it coming away in combination with the hydrogen, and burning with a purple flame, as it meets with the air. The potassium which is distilled must be condensed in naphtha kept very cold.

748. A still more productive process was pointed out by Curadau, and improved by Brunner and Wohler; potassa is prepared intimately mixed with carbon, and subjected to a strong heat in a malleable iron bottle; the carbon takes away oxygen, forming carbonic oxide gas, which is evolved, while the potassium is distilled and condensed as mentioned in the preceding paragraph. Carbon 6 & potassa 48 = carbonic oxide 14 & potassium 40. *Symb.* C&$\dot{K}$ = $\dot{C}$&K.

749. Potassium prepared by the last processes is not pure at first, being contaminated with small portions of carbon and other matters, from which it is separated by distillation.

750. The mixture of charcoal and potassa employed in Curadau's process, is procured commonly by mixing bitartrate of potassa with 1-12th of charcoal, as Brunner recommended, and heating it to redness in a crucible, taking care to exclude the air. The bitartrate contains tartaric acid, potassa, and water; and tartaric acid contains oxygen, hydrogen, and carbon. By the heat, the tartaric acid

and water are completely decomposed, the oxygen and hydrogen escaping in combination with a portion of carbon in the form of inflammable gases, excepting some carbonic acid which unites with the potassa, while the rest of the carbon remains mixed with the potassa in a minute state of division.

Compounds of Potassium.

751. I. *Compounds with Oxygen.*—POTASSA, POTASH, or OXIDE OF POTASSIUM. Symb. $\dot{K}$ = potassium 40 + oxygen 8.—Eq. 48.—Solid, white, fused at a red heat, or a little above that temperature; very deliquescent; produces great heat when combining with water; extremely corrosive. Procured by the slow oxidation of potassium.

752. With water, it forms the HYDRATE OF POTASSA; this substance is very deliquescent, soluble in water and alcohol, very corrosive, and is commonly known by the name of potassa, COMMON CAUSTIC, or POTASSA FUSA, as it is generally fused during its preparation. Symb. $\dot{H}\dot{K}$ = water 9 + potassa 48.—Eq. 57.

753. The solution of potassa turns the vegetable-blue of cabbage or violets to a fine green, but corrodes them quickly if the liquid contain much potassa. The power of potassa in corroding animal and vegetable substances appears to be connected with the property it has of promoting the union of their elements, so as to produce carbonic acid and water, with which it combines. Its taste is exceedingly acrid. For the method of ascertaining the quantity of free potassa in any solution or mixture, see Alkalimetry.

754. Potassa neutralizes the acids, and forms a great number of important salts.

755. When fused and mixed with lime, it forms the *milder common caustic.*

756. TESTS.—When potassa in solution is mixed with an excess of tartaric acid, minute white crystals of bitartrate of potassa are formed and precipitated. Perchloric acid precipitates perchlorate of potassa. Bichloride of platinum precipitates a yellow crystalline powder, consisting of chlorine, potassium, and platinum, or of muriate of potassa and platinum. Potassa gives no orange tint to the flame of the blowpipe lamp as soda does.

757. PREPARATION OF HYDRATE OF POTASSA.—Carbonate of potassa is dissolved in 10 parts of hot water, mixed with its own weight of newly slaked lime, boiled for 15 minutes, and then filtered; if kept for a day or two before it is filtered, it is unnecessary to boil it. The air must be carefully excluded during filtration, as it would communicate carbonic acid. The filtered fluid must then be evaporated to dryness in a silver or bright iron goblet, and fused by an increase of heat, when it flows like oil, and may be poured into moulds so as to assume any form that may be required. The lime retains the carbonic acid as insoluble carbonate, while the potassa remains in solution, forming the *aqua potassæ.* Carbonate of potassa 70 and lime 28 = carbonate of lime 50 and potassa 48.

758. Liebig states, that if only 4 parts of water be used with 1 of carbonate of potassa, the lime cannot separate the carbonic acid.

759. A minute quantity of carbonate of potassa is found associated with the fused potassa; by digesting it in alcohol, all the pure potassa is dissolved, and the carbonate left; the alcohol may then be separated from the pure potassa by heat.

760. Heat cannot expel the water combined with the potassa in the hydrate.

761. PEROXIDE OF POTASSIUM.—Symb. $\dddot{K}$ = potassium 40 + oxygen 24.—Eq. 64.—Solid, of an orange colour; formed when potassium is inflamed in air or oxygen, or when potassa is heated in air or oxygen. Common potassa generally contains a small quantity of the peroxide of potassium, formed during its fusion

by the absorption of oxygen from the air. It is decomposed by water, which disengages the excess of oxygen, and dissolves the potassa left.

762. II. *Compounds with Hydrogen.*—The only compounds potassium forms with hydrogen are those which have been referred to in 744.

763. III. *Compounds of Potassa with Nitric Acid, &c.*—NITRATE OF POTASSA, called also Nitre, or Salt-Petre.—Symb. :⁝N·K = nitric acid 54 + potassa 48.—Eq. 102.—Solid, transparent or white, colourless, neutral; crystallizes in 6-sided prisms; contains no water of crystallization; taste cool and penetrating. Fused at a temperature a little above 600°, when, any of the watery solution (from which it has been crystallized) that may have been lodged between the layers of the crystals is expelled. Soluble in about 3 parts of cold water, and in nearly its own weight of boiling water. A small portion is carried up mechanically with the vapour of boiling water.

764. Fused nitre, when poured into moulds so as to form small balls, is usually termed SAL-PRUNELLE.

765. Nitre is employed principally as an oxidating agent, and for the preparation of nitric acid, matches, and deflagrating mixtures.

GUNPOWDER . . .	= Nitre 75	+ Sulphur 15	+ Charcoal 10.
ARTILLERY PORT-FIRE	= Nitre 3	+ Sulphur 2	+ Gunpowder 1.
BLUE SIGNAL LIGHT	= Nitre 6	+ Sulphur 2	+ Sulphuret of Antimony 1.
WHITE SIGNAL LIGHT	= Nitre 6	+ Sulphur 2	+ Sulphuret of Arsenic 1.

766. In these and all other deflagrating mixtures containing this salt, the oxygen of the nitre is the active ingredient in promoting the rapid combustion of the inflammable matter.

767. It is also much used in warm climates for producing cold, by solution in water.—See Freezing Mixtures. Its properties as a powerful antiseptic are familiar to every one.

768. Procured from the soil in many places in India, where it is seen on the surface; water dissolves the nitre, and, by evaporation, crystals are formed. Obtained artificially by the action of air with moist animal matters and lime, by which nitric acid is procured united with lime, carbonate of potassa being added to precipitate the lime.

769. It may be prepared by neutralizing the base with the acid, but this is too expensive a method for common purposes.

770. Contaminated generally with common salt, from which it is separated by solution in hot water and crystallization, the purer crystals being deposited while the common salt is left in solution. By repeating this process with the crystals so obtained, they may at last be procured quite pure. Nitre, quite free from common salt, and dissolved in distilled water, gives no precipitate when a solution of the nitrate of silver is added; chloride of silver falls down if common salt be present.

771. NITRITE OF POTASSA.—Symb. ⁞N·K = nitrous acid 46 + potassa 48.—Eq. 94.—Said to be formed when a portion of oxygen is expelled from nitre by heat. Many regard the compound so procured as HYPONITRITE OF POTASSA mixed or combined with nitre. Sulphuric acid and water poured upon it, cause a rapid disengagement of nitrous acid and binoxide of nitrogen. When nitre is heated to a greater degree, a compound of binoxide of nitrogen and potassa is procured; and lastly, peroxide of potassium, which gives off a portion of oxygen, and becomes potassa as it is dissolved in water.

Compounds of Sulphur, Sulphuric and Sulphurous Acids, &c. with Potassium and Potassa.

772. Sulphuret of Potassium.—Symb. SK = sulphur 16 + potasssium 40. —Eq. 56.—Solid, of a dark red colour, crystalline. Dissolves in water, gives a dark green coloured solution, having a sulphureous odour; a portion of the water is believed to be decomposed, the hydrogen combining with the sulphur, and its oxygen with the potassium, hydrosulphate (hydrosulphuret) of potassa being formed.

773. Procured, 1. By heating sulphur and potassium in the required proportions; a grain or two only ought to be employed at a time, as the action is violent, and much heat and light evolved. Symb. S&K = SK. 2. By decomposing sulphate of potassa, as described in 713. 3. By the action of sulphur on potassa or its carbonate, at a high temperature. 4. By heating potassium in sulphureted hydrogen gas; the metal burns brilliantly in it, uniting with the sulphur, and disengaging hydrogen.

Decomposition of Sulphate of Potassa by Charcoal.

Fig. 26.

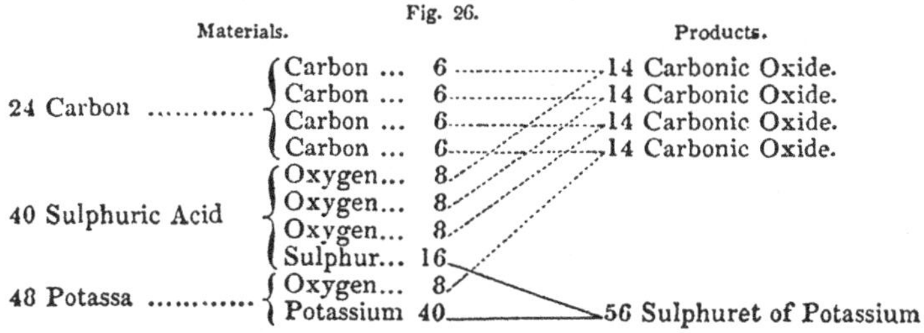

Symb. ⁝S·K&4C = SK&4·C.

774. In the 3d case, 4 eqs. of sulphur decompose 3 eqs. of potassa; 3 of sulphur take 3 of potassium from the 3 of potassa, and produce 3 of sulphuret of potassium. The other eq. of sulphur, uniting with the 3 eqs. of oxygen from the 3 eqs. of decomposed potassa, produce 1 eq. of sulphuric acid, which combines with another eq. of potassa not decomposed, sulphate of potassa being produced. Thus, there is always a portion of sulphate of potassa formed in the 3d method, and it remains mixed with the sulphuret of potassium. When carbonate of potassa is used, the carbonic acid is disengaged.

Decomposition of Potassa by Sulphur.

Fig. 27.

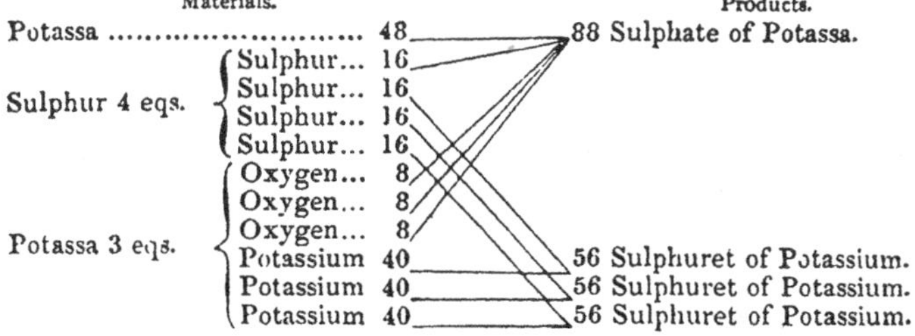

Symb. 4S&4·K = ⁝S·K&3SK.

775. It is now much doubted whether any compound of sulphur and potassa can exist; until lately, the sulphuret of potassium was always regarded as a sulphuret of potassa. Dr Dalton states, that 42 parts of potassa in solution can dissolve 56 of sulphur, and describes the compound as a QUADRISULPHURET OF POTASSA. He also mentions a TERSULPHURET OF POTASSA, and a SULPHURETED CARBONATE OF POTASSA, all prepared in solution in water. It is now the opinion of a great number of chemists that, whatever may be the quantity of sulphur taken up by potassa, or any other alkali or earth in solution, the compound never dissolves in water without decomposing a portion of it, and producing various sulphureous compounds by combining with its oxygen and hydrogen. All these solutions have an amber colour, absorb oxygen from the air, and eventually become colourless. They have a strong sulphureous odour, and an acrid bitter taste. By the addition of different acids they are decomposed.—See Hyposulphurous Acid, Bisulphureted Hydrogen, &c.

Decomposition of Sulphuret of Potassium by Water.

Fig. 28.

Materials			Products
9 Water	Hydrogen	1	17 Sulphureted Hydrogen.
	Oxygen	8	
56 Sulphuret of Potassium ...	Sulphur	16	
	Potassium	40	48 Potassa.

Symb. SK&·H = HS·K.

776. SULPHATE OF POTASSA.—Symb. ⁝S·K = sulphuric acid 40 + potassa 48. —Eq. 88.—Decrepitates by a moderate heat, and fused at a red heat. Crystals, 6-sided prisms terminated by 6-sided pyramids; small and hard, and without water of crystallization. Neutral, soluble in 15 parts of cold, and in 5 of boiling water. Taste saline and bitter.

777. Prepared in different ways from the bisulphate of potassa, which remains after the preparation of nitric acid (491). 1. On expelling the excess of acid by a strong heat. 2. By removing it with lime, which forms a very sparingly soluble compound with it. 3. By neutralizing it with potassa. Instead of using lime and potassa in the two last methods, carbonates of lime and potassa are often used, the carbonic acid being disengaged with effervescence. It may be formed by the action of sulphuric acid on potassa: this process is not economical; great heat is produced, and much sulphuric acid volatilized when the acid is strong and the potassa solid.

Fig. 29.

Materials.			Products.
128 Bisulphate of Potassa...	Potassa	48	88 Sulphate of Potassa.
	Sulphuric Acid	40	
	Sulphuric Acid	40	
50 Carbonate of lime	Carbonic Acid	22	22 Carbonic Acid.
	Lime	28	68 Sulphate of Lime.

Symb. 2⁝S+·K&:C·Ca = ⁝S·K&⁝S·Ca&:C.

From Bisulphate of Potassa and Carbonate of Potassa.—Symb. 2⁝S+·K&:C·K = 2⁝S·K&:C.

778. SULPHATE OF POTASSA WITH SULPHUR.—The name given by the Edinburgh College to a sulphureous preparation prepared by deflagrating a mixture of equal weights of sulphur and nitre in a red hot crucible. The nitric acid of the nitre is decomposed during its preparation, and sulphuret of potassium with sulphite, hyposulphate, or sulphate of potassa probably formed. It is not a mere

mixture of sulphate of potassa and sulphur, as the name may be supposed to indicate.

779. Sulphite of Potassa.—:S·K = sulphurous acid 32 + potassa 48.—Eq. 80.—Solid, uncrystallizable, effervesces from the disengagement of sulphurous acid by a stronger acid. Formed by the action of sulphurous acid with potassa, or during the partial deoxygenation of sulphate of potassa.

780. Phosphate of Potassa is prepared in the same way as phosphate of soda. Its properties are similar to those of phosphate of soda.

781. *Compounds of Carbonic Acid and Potassa.*—Carbonate of Potassa.—The subcarbonate of commerce. Symb. :C·K = carbonic acid 22 + potassa 48.—Eq. 70.—Fused by a red heat. Solid, white, not crystallizable unless with great care, deliquescent, soluble in less than its weight of water, has an acrid alkaline taste, turns the vegetable blues to a green, and turmeric to a brown.

782. Prepared from the ashes of burned wood, by pouring water upon them and boiling the solution to dryness. In this form it constitutes the potashes of commerce, a very impure carbonate of potassa, containing some caustic potassa, muriate and sulphate of potassa, sulphuret of potassium, siliceous matter, and other impurities. The Pearl Ash of commerce, so called from its whiter colour, is a purer carbonate of potassa: to prepare it potashes are dissolved and impurities allowed to subside; the solution is then evaporated to dryness, and the residue heated in a reverberatory furnace.

783. When a very pure carbonate of potassa is required, it is procured by decomposing cream of tartar (bitartrate of potassa) by heat, assisted by the action of air or of nitre. The tartaric acid is completely decomposed. Nothing remains of it but carbonic acid, formed by its carbon. When nitre is used it promotes the decomposition of the tartaric acid, and the oxygenation of its carbon, so as to produce carbonic acid; the potassa, both of the nitre and cream of tartar, remains as a carbonate; all the nitric acid is decomposed as in deflagrating processes. Pure carbonate of potassa prepared in this manner, and separated by solution and evaporation from adhering impurities, is called Salt of Tartar.

784. The White Flux is a pure carbonate of potassa prepared by deflagrating equal weights of cream of tartar and nitre. Black Flux is composed of carbonate of potassa mixed with carbon; and is procured by deflagrating 2 parts of cream of tartar with one of nitre; the black colour depends upon an excess of carbon derived from the tartaric acid, as there is not a sufficient quantity of oxygen in the nitric acid of the nitre used, to convert it all into carbonic acid.

785. Bicarbonate of Potassa, commonly called Carbonate or Super-carbonate.—Symb. 2:C+·K, and when Cr. ·H(2:C+·K) = water 9 + carbonic acid 44 + potassa 48.—Eq. 101.—Solid, crystalline, forming 8-sided prisms, less acrid than the carbonate, does not deliquesce, loses 1 eq. of carbonic acid by boiling in water, or by exposure to a red heat. Turns the colouring matter of the cabbage to a green, but feebly, compared with the carbonate. Tartaric acid 75 grs., or citric acid 76, decompose 101 of the bicarbonate, disengaging all the carbonic acid; 30 or 40 grains of tartaric or citric acid, dissolved in a few ounces of water, and added to a corresponding quantity of bicarbonate of potassa, also in solution, produce a good effervescing draught. An excess of acid is generally used, as it is more palatable than an excess of alkali.

786. Prepared by exposing the carbonate either dry or in solution to carbonic acid gas.

787. *Compounds of Potassa with Vegetable Acids.*—Acetate of Potassa.—Eq. 99.—Solid, crystalline, presents a foliaceous appearance, fused by a moderate

heat, decomposed rapidly by a red heat, neutral, deliquescent, very soluble in water.

788. Procured by neutralizing potassa in solution with acetic acid and evaporating the solution cautiously till it assume a syrupy consistence. Carbonate of potassa is commonly used instead of the pure potassa, carbonic acid gas being then disengaged with effervescence.

789. TARTRATE OF POTASSA is very soluble in water, and may be made by neutralizing bitartrate of potassa with potassa or its carbonate, or by removing the excess of tartaric acid in the bitartrate by chalk. When carbonate of potassa is used, Symb. 2*T*+·K&:C·K = 2*T*·K&:C. See Tartaric Acid.

790. BITARTRATE OF POTASSA, or Cream of Tartar, is obtained from wine casks, where it has been deposited slowly from the juice of the grape. It is soluble in 70 of cold and 15 of hot water. It is also highly acid.

791. *Compounds of Chlorine with Potassium and Chloric Acid with Potassa.*—CHLORIDE OF POTASSIUM.—Symb. ClK = chlorine 36 + potassium 40.—Eq. 76.—May be formed by the action of potassium on chlorine, which burns brilliantly in this gas. It may also be formed by muriatic acid and potassa, the chlorine combining with the metal, and the hydrogen with the oxygen of the potassa. It is usually known in commerce by the name of the MURIATE OF POTASSA. It is prepared principally by indirect processes, and is similar to chloride of sodium.

792. CHLORATE OF POTASSA.—⁚⁚Cl·K=chloric acid 76 + potassa 48.—Eq. 124. It is formed by the action of chlorine on a solution of potassa, in the manner described in 639. It is transparent and colourless, crystallizes in tables with 4 or 6 sides, gives off oxygen by the action of heat, and leaves chloride of potassium; phosphorus, sulphur, and a number of inflammable substances detonate when mixed with it on striking them with a hammer. Mixed with half its weight of sugar, it forms a very inflammable composition, which inflames rapidly on touching it with a hot iron rod, or with a drop of cold sulphuric acid, peroxide of chlorine being evolved in the latter case, which inflames immediately the whole of the mixture.

793. *Compounds of Iodine, &c. with Potassium and Potassa.*—IODIDE OF POTASSIUM.—Symb. IK = iodine 124 + potassium 40.—Eq. 164.—Solid, crystallizes in cubes, soluble in water and alcohol. Deliquesces in a moist atmosphere. Prepared by bringing iodine and potassium in contact; they combine rapidly when gently heated, with evolution of heat and light, I&K = IK: or by evaporating to dryness a solution of the hydriodate of potassa, the hydrogen of the acid combining with the oxygen of the potassa, and leaving the iodide of potassium.

Fig. 30.

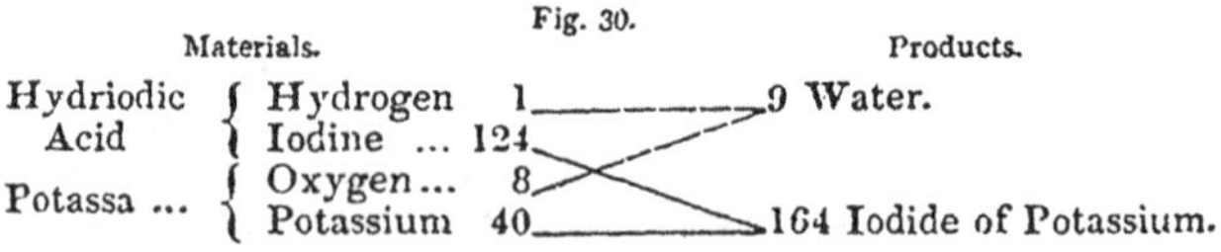

Symb. HI&·K = ·H&IK.

If iodine be treated with a solution of potassa, water is decomposed, as in 639, and iodate and hydriodate of potassa are procured. Water 5 eqs. & iodine 6 eqs. & potassa 6 eqs. = iodate of potassa & 5 eqs. hydriodate of potassa.—Symb. 5·H&6·K&6I = ⁚⁚I·K&5HI·K. By evaporating to dryness, and heating the product, nothing is left but iodide of potassium.—Symb. ⁚⁚I·K&5HI·K = 6IK&5·H&O^6 This process was pointed out by Dr W. Gregory.

794. Another process for preparing this compound consists in boiling iodine with

metallic iron or zinc in water, when a compound of a metallic oxide with hydriodic acid is procured in solution, part of the water being decomposed. On adding carbonate of potassa, carbonate of iron or of zinc is precipitated, and hydriodate of potassa is obtained in solution, from which the iodide is to be obtained by evaporation. Symb. HI·K = ·H&IK.

Preparation of the Hydriodate of Iron.

Fig. 31.

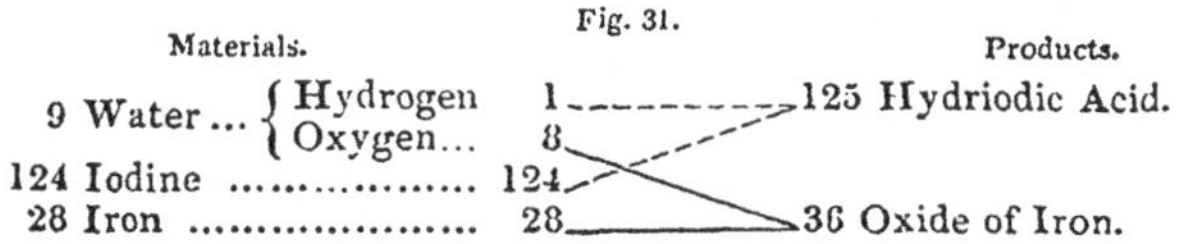

Symb. ·H&Fe&I = HI+·Fe.

Decomposition of the Hydriodate of Iron by Carbonate of Potassa.

Fig. 32.

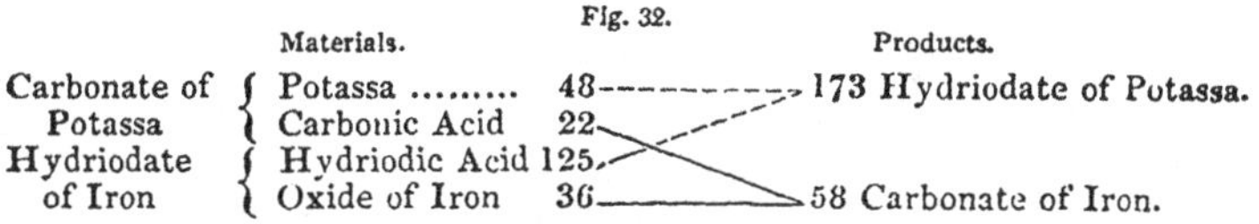

Symb. :C·K&HI·Fe = HI·K&:C·Fe.

795. Iodide of potassium is often much adulterated with carbonate of potassa; the quantity present may be ascertained by finding the quantity of sulphuric acid necessary to neutralize it; 40 of sulphuric acid indicate 70 of carbonate of potassa.

CHAP. II. SODIUM.

796. Symb. Na. from Natron, the old name for carbonate of soda, one of its most important compounds.—Eq. 24.—Sp. gr. 0.972.—Soft and malleable at 32°, fused at 200°, and volatilized at a full red heat. Decomposes in water with rapidity, but does not inflame so readily as potassium. When moistened with water on charcoal, wood, or some other bad conductor, or with a solution of gum in water, it always inflames. It resembles potassium much in its appearance, and in all its chemical relations, but does not attract bodies so powerfully, so that potassium can detach it from all its combinations.

Compounds of Sodium.

797. *Compounds with Oxygen,*—SODA, or OXIDE OF SODIUM.—Symb. ·Na = sodium 24 + oxygen 8.—Eq. 32.

798. HYDRATE OF SODA.—Symb. ·H·Na.—May be procured from carbonate of soda in the same manner as hydrate of potassa from its carbonate. Similar to hydrate of potassa in appearance and properties. Carbonate of soda 54 and lime 28 = carbonate of lime 50 and soda 32.—Symb. :C·Na&·Ca = :C·Ca&·Na.

799. TESTS.—Gives an orange tint in the flame of the blowpipe, and is not precipitated by tartaric acid, perchloric acid, or bichloride of platinum.

800. PEROXIDE OF SODIUM.—Symb. ⁝Na^2 = sodium (2 eqs.) 48 + oxygen (3 eqs.) 24.—Eq. 72.—Orange colour, decomposed by water in the same manner as peroxide of potassium, the oxygen in excess escaping and soda being left in solution.

801. *Soda and Nitric Acid.*—NITRATE OF SODA.—Symb. :⁝N·K = nitric acid

54 + soda 32.—Eq. 86.—Resembles common nitre very much, but is not so valuable on account of its deliquescence. It is now largely used in some manufactories for the preparation of nitric acid. It is imported from South America.

802. *Compounds of Sulphur and its combinations with Sodium and Soda.*—SULPHURET OF SODIUM, HYDROSULPHURET OF SODA, are so similar, in general, to the corresponding compounds of potassium, that it is unnecessary to detail their properties, unless where some marked difference presents itself. *Preparation of Sulphuret of Sodium.*—Sulphate of soda 72 and carbon 24 = sulphuret of sodium 40 and carbonic oxide 56. Symb. ⁝S·Na&4C = SNa&4·C.

803. SULPHATE OF SODA.—Symb. ⁝S·Na = sulphuric acid 40 + soda 32, and when crystallized the Symb. is 10·H + (⁝S·Na) = water 90 + sulphate of soda 72. Commonly called GLAUBER'S SALT.—Crystallizes in 6-sided prisms, which are soluble in 3 parts of cold water, and in 1–3d of their weight of water at 90°; at a higher temperature the solubility of the salt diminishes. A saturated hot solution allowed to cool in a calm and still atmosphere, or with the air altogether excluded, does not in general crystallize on cooling, but whenever it is agitated, or a small piece of crystal introduced, the whole of it immediately passes to the solid form in beautiful radiations.

804. Sulphate of soda is efflorescent, soon acquiring a white appearance, and falling down in powder. It is extremely bitter.

805. It is procured in general by the action of aqueous sulphuric acid on common salt, during the preparation of aqueous muriatic acid. Any excess of sulphuric acid may be withdrawn in the same manner as from the supersulphate of potassa (777).

806. Cr. PHOSPHATE OF SODA.—Symb. ·H + (:⁝P^2+2·Na) = phosphoric acid 72 + soda 64 + water 9.—Eq. 143.—Soluble in 4 parts of cold water, and in nearly 2 parts of boiling water. By heat it is converted into the pyrophosphate of soda of Professor Clarke.

807. Prepared by the action of superphosphate of lime in solution on carbonate of soda. The phosphoric acid, uniting with the soda, disengages carbonic acid, and forms the phosphate of soda; the phosphate of lime having lost its excess of phosphoric acid is precipitated. This is separated by filtration, and the solution of phosphate of soda is crystallized by evaporation. A slight excess of carbonate of soda favours the crystallization. The solution of superphosphate of lime contains a little sulphate of lime; this, acting with the carbonate of soda, produces sulphate of soda and carbonate of lime. The sulphate of soda remains with the phosphate in solution, and is not easily separated; the carbonate of lime is precipitated.

808. *Compounds of Carbonic Acid with Soda.*—CARBONATE OF SODA.—Symb. :C·Na; and when crystallized 10·H + (:C+·Na).—The quantity of water, however, varies according to the temperature at which the carbonate is crystallized.

809. The crystals are efflorescent, and lose their water of crystallization quickly by heat, after which the dry carbonate may be fused. Soluble in about 2 parts of cold water, and in rather less than their weight of boiling water. Taste acrid and alkaline; turns the colouring matter of the cabbage to a green; decomposed rapidly with effervescence by the stronger acids.

810. Prepared from KELP or BARILLA, the ash that remains after burning seaware. Water is poured upon it to dissolve the carbonate of soda, and the solution crystallized by evaporation. It is manufactured also in large quantity by heating carbonate of lime, charcoal, and sulphate of soda; the charcoal tends to produce sulphuret of sodium, and the lime to remove sulphur or sulphuric acid. The carbonic acid of the carbonate of lime, and that from the flame of the furnace made

to pass over the soda, enable it to acquire carbonic acid. By solution and evaporation, it is procured with facility in crystals.

Preparation of Carbonate of Soda by Carbonate of Potassa.—Sulphate of soda 72 and carbonate of potassa 70 = sulphate of potassa 88 and carbonate of soda 54. The less soluble sulphate of potassa is precipitated. Symb. ⁝S·Na&:C·K = ⁝S·K&:C·Na.

811. BICARBONATE OF SODA.—Symb. 2:C+·Na.—When crystallized, it contains, according to some chemists, 1 eq. of water = 9. Its properties are similar to those of the bicarbonate of potassa, and it is prepared in the same manner.

812. TARTRATE OF POTASSA AND SODA, or ROCHELLE SALT, is composed of 1 eq. of tartrate of potassa 114 + 1 eq. of tartrate of soda 98 + 8 eq. water 72 = 284. It is soluble in 5 parts of water, and gives prismatic crystals.

813. Prepared on neutralizing bitartrate of potassa, mixed with boiling water, by carbonate of soda; the carbonic acid of the carbonate escapes.

Fig. 33.

Materials Decomposed.				Products.
Carbonate of Soda	Carbonic Acid	22	22	Carbonic Acid.
	Soda	32	98	Tartrate of Potassa and Soda.
Bitartrate of Potassa	Tartaric Acid	66		
	Tartaric Acid	66		
	Potassa	48	114	

Symb. 2*T*+·K&:C·Na = 2*T*+·K+·Na&:C.

814. CR. BIBORATE OF SODA, or BORAX, often called SUB-BORATE OF SODA, as the qualities of the alkali predominate, though there are 2 eqs. of acid combined with it. Crystallizable, soluble in 6 parts of boiling and 20 of cold water. By heat its water of crystallization is expelled, and a transparent glass formed, which is much used as a flux. Consists of 2 eqs. boracic acid 48, 1 eq. soda 32, and 10 eqs. water 90, = 180.

815. *Compounds with Chlorine, &c.*—CHLORIDE OF SODIUM.—Symb. ClNa = chlorine 36 + sodium 24.—Eq. 60.—Solid, crystallizes in cubes and in hollow pyramids with 4 sides; contains no water of crystallization, decrepitates, and then fuses by heat. The decrepitation depends upon the expansion of water lodged mechanically between the plates of the crystals. Soluble in 2½ parts of cold water, and requires nearly the same quantity of boiling water. When it decomposes water, chloride of water 60 and water 9 = muriate of soda 69. Symb. ClNa&·H = HCl+·Na.

816. Procured in its purest form from ROCK SALT, which, when mixed with earthy matter, may be separated by solution and crystallization. BAY SALT comes next in purity, and is prepared by the spontaneous evaporation of sea-water in warm climates. Salt obtained by the rapid boiling down of sea-water in a pan or boiler is the least pure of all the varieties of common salt met with in commerce, containing usually a notable quantity of magnesian salts, principally muriate of magnesia, which renders it deliquescent, and prevents it from being employed with so much advantage as an antiseptic.

817. Carbonate of soda in solution, added to a pure solution of common salt, gives no precipitate; where magnesia is in solution, the liquid is immediately rendered turbid, and all the magnesia is precipitated in the form of a carbonate by boiling.

818. CHLORO-CARBONATE OF SODA is a compound much employed in solution for removing noxious odours, by decomposing animal and vegetable effluvia. It is procured by transmitting chlorine gas through a solution of carbonate of soda. It is familiarly known by the name of Labarraque's disinfecting fluid. The chlo-

rine does not at first decompose the carbonate of soda, but by long keeping chlorate and muriate of soda are formed, as in 639.

CHAP. III. LITHIUM.

819. Lithium is the metallic base of lithia, which consists of lithium and oxygen.—Symb. L.—Eq. of lithium 10, of lithia 18, containing 8 of oxygen. It was discovered by Arfvedson in 1818. Lithia possesses all the qualities of an alkali, and is distinguished from potassa and soda by the very sparing solubility of its phosphate and carbonate.

CHAP. IV. AMMONIA.

820. The history of ammonia is inserted here, though it contains no metallic matter in its composition, as it is very similar to the alkalis in all its properties and combinations, and as the various reactions connected with it will be better understood in this place than they could previously have been.

821. Symb. NH^3 = nitrogen 14 + hydrogen 3.—Eq. 17 by W, ☐☐ by V.—Sp. gr. 0.59.—W of 100 c. i. 18 grs.—Condensed into a liquid by a pressure of 6.5 atmospheres at 50°. Decomposed when transmitted through red hot tubes, or when the electric spark is passed through it for some time, 2 volumes of ammoniacal gas being resolved into 3 of hydrogen and 1 of nitrogen.

822. Gaseous, transparent, colourless, very pungent and peculiar odour, cannot support respiration or combustion. Inflammable, but can scarcely be made to burn with air; 2 volumes mixed with 1½ of oxygen detonate violently on applying a lighted match.

Fig. 34.

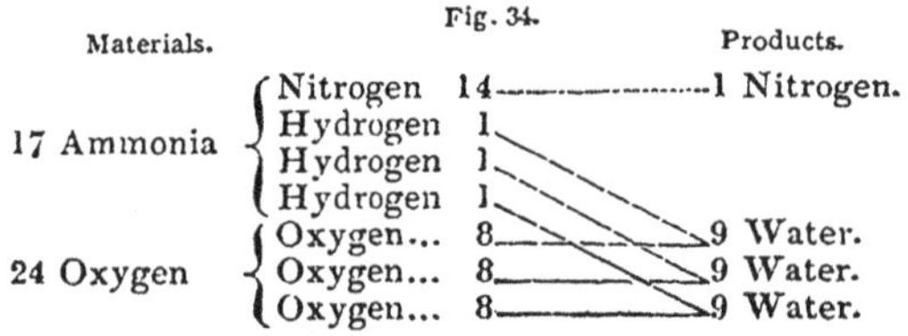

Symb. $NH^3\&O^3$ = N&3·H.

823. Absorbed with explosive rapidity by water, producing the liquid usually termed Ammonia, or Spirit of Hartshorn, being prepared in general from the horns of animals, and from other kinds of animal matter. The solution varies in its density, which is less and less, the larger the quantity of ammonia combined with the water. Dr Thomson states that water can take up 780 times its bulk of ammoniacal gas. The following table shews the density of aqueous ammonia of different degrees of strength:

Specific gravity.	Ammonia in 100 parts.	Specific gravity.	Ammonia in 100 parts.	Specific Gravity.	Ammonia in 100 parts.	Specific Gravity.	Ammonia in 100 parts.
.8550	32.50	.9166	22.07	.9435	14.53	.9573	10.82
.8875	29.25	.9255	19.54	.9476	13.54	.9577	10.17
.9000	26.00	.9326	17.52	.9513	12.40	.9619	9.60
.9045	25.37	.9385	15.88	.9545	11.56	.9672	9.50

824. With alcohol it combines readily, forming what is usually termed Alcohol Ammoniatum, or Ammoniated Alcohol. It is prepared by passing ammoniacal gas into alcohol, which must be kept cool.

825. Tests.—Ammonia is easily recognised in solution by the very deep and brilliant blue colour which it gives to solutions containing copper (see Copper). When combined with an acid, it may be disengaged by potassa, soda, or lime, which retain the acid previously combined with it, while a gentle heat causes the ammonia to be expelled in the gaseous form. In this condition it may be recognised by the white fumes it produces with nitric, carbonic, or muriatic acids, and by its alkaline reaction on turmeric paper, which it turns to a brown, and on cabbage paper, which becomes green; a very slight heat is sufficient to restore the original colour of the test paper, the ammonia being expelled.

826. Ammonia has all the characteristic properties of an alkali, though differing so widely from the common alkalis in its composition. It has a very acrid alkaline taste; corrodes animal and vegetable matters; affects the blue colouring matter of cabbage and violets, and also turmeric, in the same manner as potassa; neutralizes the acids and forms salts; these, however, are either decomposed or volatilized by a red heat. From its volatility it is often termed the volatile alkali; potassa, soda, and lithia, being denominated fixed alkalis.

827. Preparation.—Ammonia is procured from animal substances; these are subjected to a red heat, and the gas evolved is condensed in water. The proximate principles of the animal kingdom are almost all composed of carbon, oxygen, hydrogen, and nitrogen. Subjected to heat when the air is excluded, their elements assume a new arrangement, and part of their hydrogen and nitrogen combining together, ammonia is the product, chiefly in union with carbonic acid. By combining it with muriatic acid, and separating it again by means of potassa. soda, or lime, it is procured free from any empyreumatic odour, and in the condition in which it is condensed by water to afford the purest ammonia. In conducting this part of the process, muriate of ammonia is generally mixed with slaked lime, and the ammoniacal gas disengaged by the action of heat is condensed in water kept very cold; the nature of the action is particularly explained in the diagram page 4, where the proportion of materials that affect each other, and of the products formed, is pointed out. *Symb.* $HCl+NH^3\&\cdot Ca = ClCa\&\cdot H\&NH^3$.

Compounds of Ammonia.

828. Nitrate of Ammonia.—Symb. $:\!\vdots N+NH^3$ = nitric acid 54 + ammonia 17. Eq. 71.—Symb. of Cr. Nitrate of Ammonia = $\cdot H+(:\!\vdots N+NH^3$ = water 9 + nitrate of ammonia 71 =. Eq. 80.

829. Very deliquescent, produces great cold during solution in water, crystallizes in prisms, when evaporated at a temperature not exceeding 100°; a fibrous mass is procured if it be evaporated at 212°; and when fused by heat continued so that it acquires a temperature of 300° or upwards, it becomes quite compact on cooling.

830. Its decomposition, so as to produce nitrous oxide, has been already noticed, par. 470; when subjected suddenly to a red heat, it deflagrates with rapidity, the hydrogen of the ammonia uniting with part of the oxygen of the nitric acid, and forming water.

831. Sulphate of Ammonia.—Symb. $\vdots S+NH^3$ = sulphuric acid 40 + ammonia 17. Eq. 57.—Prepared easily by neutralizing dilute sulphuric acid with ammonia or its carbonate, carbonic acid being disengaged when the carbonate is used. It is found in soot, associated with sulphite of ammonia, the sulphur of the acid being derived from the iron-pyrites so frequently met with in coal, and the ammonia formed by the decomposition of the bituminous matter. Soluble in water, crystallizes in 6-sided prisms.

832. Hydrosulphate or Hydrosulphuret of Ammonia.—$HS+NH^3$ =

hydrosulphuric acid 17 + ammonia 17. Eq. 34.—Procured in the solid form when sulphureted hydrogen and ammoniacal gases are brought in contact. Generally prepared in solution by passing sulphureted hydrogen gas into a solution containing ammonia. The solution is much employed as a test for metals, containing a large quantity of sulphureted hydrogen in a condensed form, and in such a condition that it acts freely upon them, the ammonia removing entirely any acid matter which might otherwise interfere with its operation.

833. PHOSPHATE OF AMMONIA may be prepared in the same manner as phosphate of soda, or by directly neutralizing ammonia with phosphoric acid. Exposed to a red heat it is completely decomposed, glacial phosphoric acid remains, all the ammonia is expelled.

834. PHOSPHATE OF SODA AND AMMONIA, MICROCOSMIC SALT, or SALT OF PHOSPHORUS, is prepared by dissolving 1 equivalent of the common phosphate of soda and 1 of muriate of ammonia in no more boiling water than is necessary, and allowing the liquid to cool. Crystals are deposited containing phosphoric acid, soda, ammonia, and water. In experiments with the blowpipe, this salt is extremely useful, as it is easily melted by heat, and the water and ammonia expelled, so that a very fusible biphosphate of soda remains, which acts powerfully both as an acid compound and as a flux.

835. CARBONATE OF AMMONIA may be procured by mixing 2 measures of ammonia and 1 of carbonic acid at the mercurial trough; they condense immediately into a solid white powder.

836. BICARBONATE OF AMMONIA may be prepared by passing a stream of carbonic acid through a solution of the common carbonate of ammonia, till it ceases to absorb any more, and evaporating the fluid with great care by a moderate heat.

837. SESQUICARBONATE OF AMMONIA.—Symb. $\cdot H + (3{:}C + 2NH^3)$ = water 9 + carbonic acid 66 + ammonia 34, or water 9 + carbonate of ammonia 39 + bicarbonate of ammonia 61 = eq. 109. This is the most important of the carbonates of ammonia; it is commonly called SUBCARBONATE OF AMMONIA. It is solid, has a strong and pungent ammoniacal odour, volatilizes slowly at natural temperatures, and is converted rapidly into vapour by heat. By exposure to the air, the ammonia escapes more rapidly than the carbonic acid, so that it is apt to pass to the state of bicarbonate of ammonia, losing its penetrating odour, transparency, and hardness, and becoming a brittle porous mass. It is very soluble in water, and rises rapidly with its vapour when boiled.

838. Prepared from muriate of ammonia by carbonate of lime, in the same manner that uncombined ammonia is prepared from the muriate by lime; see 29. The only difference in the nature of the reaction, arises from the circumstance that there is carbonic acid associated with the lime used to decompose the muriate, which attaches itself to the ammonia evolved. Were the temperature necessary for the decomposition less elevated, all the ammonia disengaged would combine with the carbonic acid, and a simple carbonate of ammonia would be produced;

Fig. 35.

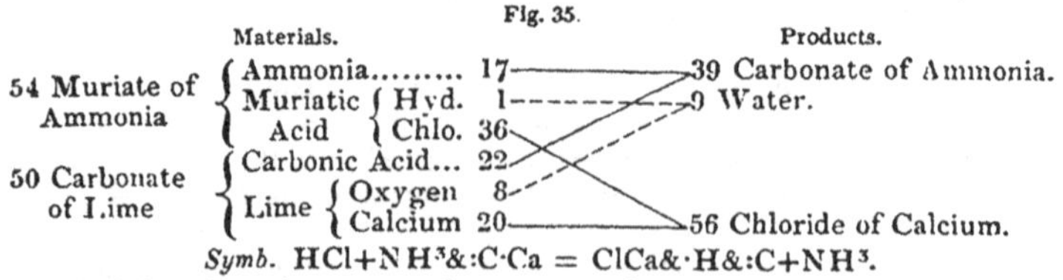

Symb. $HCl + NH^3 \& {:}C \cdot Ca = ClCa \& \cdot H \& {:}C + NH^3$.

instead of this, from the escape of a portion of ammonia, it is only a sesquicarbonate that is formed. The diagram explains the preparation of the carbonate.

839. Dried carbonate of soda decomposes muriate of ammonia with greater facility than carbonate of lime.

840. Acetate of Ammonia is prepared by saturating acetic acid diluted with water by carbonate of ammonia, the carbonic acid escaping with effervescence; it is usually called the *Spirit of Mindererus*, and is employed medicinally. It is very deliquescent and soluble. Acetic acid 51 and ammonia 17 = acetate of ammonia 68.

841. Oxalate of Ammonia is prepared in a similar manner, and is very easily crystallized. It is much used as a test of lime. Oxalic acid 36 and ammonia 17 = 53 oxalate of ammonia. Symb. $\colon C^2\&NH^3 = \colon C^2+NH^3$.

842. Chlorine decomposes ammoniacal gas with the evolution of heat and light, accompanied by the production of a lambent flame. Chlorine 3 eqs. decompose 1 eq. of ammonia; each eq. of chlorine combines with an eq. of hydrogen, which it takes from the ammonia, so that 3 eqs. of muriatic acid are formed, and 1 of nitrogen disengaged. Each eq. of muriatic acid then combines with an eq. of ammonia not decomposed, producing muriate of ammonia. These constitute the white vapours that always appear when an excess of ammonia comes in contact with chlorine. If a stream of chlorine gas be passed into a solution of ammonia in water, similar effects are produced, and pure nitrogen may be collected as it is evolved.

Fig. 36.

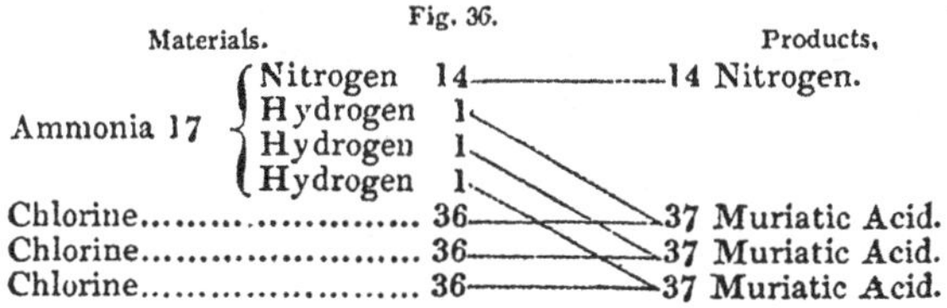

Symb. $NH^3\&3Cl = 3HCl\&N$.

843. Muriate of Ammonia or Sal-Ammoniac.—Symb. $HCl+NH^3$ = muriatic acid 37 + ammonia 17. Eq. 54. Solid, crystalline, presents a fibrous texture when sublimed, and forms feathery-looking crystals when deposited from its solution in water. Taste cool, acrid, and penetrating. Soluble in rather less than 3 parts of cold water, and in its own weight of boiling water. Alcohol dissolves a small quantity of it. At 300° it sublimes slowly.

844. Prepared in different ways: I. By saturating the impure carbonate of ammonia contained in the watery gas liquor procured from coal, when a solution of muriate of ammonia is obtained, which gives crystals on evaporation. These must be purified by sublimation. II. By heating animal matters impregnated with a solution of muriate of magnesia, the muriatic acid unites with ammonia produced during the decomposition of the animal matter, is slowly sublimed, and the magnesia is left. III. By the mutual decomposition of the sulphate of ammonia in soot and common salt; muriate of soda is formed as the common salt acts on a portion of water, and this decomposing the sulphate of ammonia gives sulphate of soda and muriate of ammonia. By crystallization and sublimation the muriate is obtained pure.

ORDER II. TERRIGENOUS METALS.

I. ALKALINE TERRIGENOUS METALS.

845. These metals, united with oxygen, form the alkaline earths, Lime, Baryta, Strontia, and Magnesia; which, with the exception of magnesia, corrode animal and vegetable substances in the same manner as the alkalis, but much more feebly. Magnesia, compared with the three preceding earths, is so very bland and inert, that, unless under particular circumstances, it cannot be said to have any very caustic properties. All these earths neutralize the acids, turn the cabbage blue to a green, and turmeric to a brown, in the same manner as the alkalis. Their carbonates are very insoluble, but their bicarbonates are dissolved readily when they are in a minute state of division, and mixed with a large quantity of water. Their sulphates are either insoluble, or very sparingly soluble, in water, excepting the sulphate of magnesia, which is a very soluble salt.

846. The metals which belong to this order have not been so accurately examined, nor can they be procured with so much facility as those of the preceding order. Calcium, the base of lime, is procured by the action of the galvanic current on carbonate of lime moistened with a little water, placing it on a platinum tray, and making a little cup or cavity in its surface, that it may hold a globule of mercury. The platinum tray having been connected with the positive pole of the battery, and the negative pole introduced into the mercury, a small portion of lime is decomposed, its oxygen being removed to the positive pole, while the calcium combines with the mercury. By heat the mercury is distilled, and the calcium procured by itself. Barium, strontium, and magnesium, have been procured in a similar manner.

847. Peroxides of the greater number of these metals have been described, which, with the exception of the peroxide of barium, have not been applied to any important purpose. They may be formed by passing oxygen over the heated oxide till it ceases to absorb any more; or, by heating the oxide to dull redness and adding four parts of chlorate of potassa; it communicates the oxygen required and leaves chloride of potassium, which may afterwards be removed by cold water.

CHAP. I. CALCIUM.

848. Symb. Ca. Eq. 20.—Has a white colour and metallic lustre, oxidated on heating in the air, and converted into lime.

849. Lime, Quicklime, or Oxide of Calcium.—Symb. ·Ca = calcium 20 + oxygen 8. Sp. gr. 2.3.—Extremely infusible, but is rendered semivitreous, and volatilized slowly by the oxyhydrogen blowpipe, condensing and forming minute crystals, or collecting as a white smoke a little way from the place where it is converted into vapour. Gives out a brilliant light when intensely heated.

850. White, friable; has a harsh, styptic, and alkaline taste. Combines with water, and forms HYDRATE OF LIME, commonly called SLAKED LIME, great heat being evolved; it is composed of lime 28 + water 9. Eq. 37.

851. Lime Water.—Lime is sparingly soluble in water; according to Dalton, 778 of water are required to dissolve 1 of lime. It is less soluble in hot water than in cold, and hence a cold saturated solution of lime in water always becomes turbid on heating it, from the deposition of lime. It is prepared merely by agi-

tating lime in a bottle with water, and allowing it to stand till the excess of lime employed shall have subsided. If lime-water be required quickly, it may be filtered to separate undissolved lime, care being taken to exclude the air, which would communicate carbonic acid and precipitate the lime.

852. Lime-water is perfectly colourless, but has a brilliant transparency, which is not observed in any other kind of water. Evaporated in the exhausted receiver of the air-pump, with the aid of the sulphuric acid (212), the hydrate crystallizes in hexahedrons. Its taste is similar to that of lime, and it affects the vegetable colouring matters in the same way.

853. The salts of lime communicate a reddish tint to alcohol and some other inflammable fluids during combustion, when they have been previously dissolved or mixed with them.

854. Lime is precipitated from solutions of its salts by the alkalis potassa and soda, but not by ammonia. Alkaline carbonates give a white earthy precipitate of carbonate of lime when mixed with solutions of its salts, the alkali of the carbonate uniting with the acid previously combined with the lime. Sulphuric acid and concentrated solutions of sulphates give copious precipitates with strong solutions of salts of lime, but none if they are diluted much with water, sulphate of lime, the compound precipitated when there is much lime in solution, being more soluble in water than lime itself. Oxalic acid is the most delicate test of lime, forming a very insoluble compound with it: when the lime is combined with an acid, oxalate of ammonia is preferred to oxalic acid alone, as the ammonia unites with the acid, and allows the oxalic acid to act without any impediment upon the lime. If the salt of lime contain an excess of acid, the excess ought to be carefully neutralized by ammonia.

855. *Lime is distinguished from alkalis* by the insolubility of its carbonates. *From baryta and strontia*, by its solution in water giving no precipitate with sulphuric acid, which baryta and strontia do. Its soluble salts, also, diluted with as much water to the lime as exists in lime-water, give no precipitate with this acid. *From magnesia*, by its solubility in water, and the comparative insolubility of its sulphate. Magnesia is so insoluble in water that it communicates little or no sensible properties to it; but its sulphate is very soluble in water, while only a small quantity of sulphate of lime is dissolved by this fluid.

856. Preparation of Lime.—From limestone, chalk, marble, or calcareous spar, all of which consist principally, if not entirely, of carbonate of lime. Calcareous spar and white marble are the purest carbonates, containing, in many cases, only a fraction per cent. of impurities. From all these the carbonic acid is expelled by heat, and the limestone or chalk is said to be burned, in common language, but the only change which is effected in it is the expulsion of the carbonic acid. Carbonate of lime 50 = carbonic acid 22 and 28 lime. Symb. :C·Ca = :C&·Ca. On the large scale, alternate beds of limestone and fuel are placed in the furnace or lime-kiln; on the smaller scale, the carbonate to be decomposed may be put into a crucible. When perfectly free from carbonic acid, and diffused through water, it gives no effervescence on adding muriatic acid.

857. Lime free from carbonic acid absorbs water with rapidity. If not kept carefully from the air it soon attracts moisture and carbonic acid, returning to the state of carbonate.

858. The setting of lime in common mortar depends upon its attracting carbonic acid from the air, through the porous mass composed of lime and sand. When lime or mortar has been kept for a long time exposed to the air before it is used for building, it is said to be spent, as it has already become charged with carbonic acid, and cannot set or act so well as a cement.

859. PEROXIDE OF CALCIUM.—Symb. :Ca. Eq. 36 = calcium 20 + oxygen 16 —See 847.

Compounds of Calcium and Lime.

860. NITRATE OF LIME.—Symb. ::N·Ca = nitric acid 54 + lime 28. Eq. 82. Deliquescent. Soluble in alcohol, which separates it from nitrates of baryta and strontia. Formed by neutralizing nitric acid diluted with water by lime. The carbonate may be employed, which dissolves with effervescence, the carbonic acid escaping.

861. *Compounds with Sulphur, Sulphuric Acid, &c.*—SULPHURET OF CALCIUM.—Symb. SCa = sulphur 16 + calcium 20. Eq. 36.—Soluble in water, and believed to decompose it (711), at the same time producing hydrosulphuret (hydrosulphate) of lime. Employed principally for the preparation of sulphureted hydrogen. Prepared by decomposing sulphate of lime mixed with charcoal, heating it in a crucible. Symb. ⁝S·Ca&4C = SCa&4·C.—See Decomposition of Sulphates, 713. When well prepared, it effervesces rapidly in diluted muriatic acid, producing abundance of sulphureted hydrogen gas.

862. QUADRISULPHURET OF LIME.—Symb. S^4·Ca = sulphur 64 + lime 28. Eq. 92.—The compound formed when sulphur, lime, and water, are boiled together, according to the opinion of Dalton, whether an excess of lime or sulphur be employed, and communicating to it the amber tint which it presents. Very different opinions have been entertained as to the manner in which the sulphur exists in the solution; it is often termed SULPHURETED HYDROSULPHURET OF LIME, from the opinion that bisulphureted hydrogen exists in it, and communicates its peculiar properties, though it is obvious, that when this compound is produced by one portion of sulphur acting on the hydrogen of the water, another portion must combine at the same moment with the oxygen, as none is disengaged.

863. This compound has been employed principally in removing oxygen from the air in eudiometrical experiments; Dr Hope's eudiometer is best adapted for this purpose, and consists of a bottle intended to contain the liquid, and a graduated tube for containing the air to be examined, which is fitted to the mouth. As the absorption of oxygen goes on, the bottle is opened from time to time under water, by taking out a stopple at the side. It is also employed for the preparation of precipitated sulphur. The compound prepared with sulphur, potassa, and water, is often substituted for it.

864. SULPHATE OF LIME.—Symb. ⁝S·Ca = sulphuric acid 40 + lime 28. Solid, tasteless, infusible. Soluble in about 500 parts of cold water, and a little more soluble in boiling water, *hard water* being produced. Mixed with water, after it has been reduced to powder, so as to form a thick cream, it gradually consolidates into a hard and firm mass, the water combining intimately with it; in this condition it forms PLASTER-OF-PARIS, so much employed in taking casts. Sulphate of lime is often precipitated by the action of sulphuric acid or sulphates on salts of lime. Sulphuric acid and muriate of lime = sulphate of lime and muriatic acid. Symb. ⁝S&HCl+·Ca = ⁝S·Ca&HCl. Sulphate of soda and muriate of lime = sulphate of lime and muriate of soda. Symb. ⁝S·Na&HCl·Ca = ⁝S·Ca&HCl·Na.

865. It occurs abundantly in nature, and forms GYPSUM, SELENITE, and ALABASTER, all of which contain water of crystallization, and this must be expelled before it can be used for making plaster-of-Paris.

866. PHOSPHATE OF LIME.—Found in large quantity in the bones of animals, from which it is procured by burning off the animal matter. White, earthy, infusible. Soluble in diluted nitric and muriatic acids, and precipitated again when

the acid which had dissolved it is neutralized. It is used principally for the preparation of phosphorus, phosphoric acid and its salts, and in making cupels.—See Cupellation. Many SUPERPHOSPHATES OF LIME have been described, all soluble in water.

867. PHOSPHURET OF CALCIUM.—Procured most conveniently when a piece of phosphorus is thrown upon lime heated to redness in a crucible, a cover being immediately put upon it, and a quantity of sand thrown upon the cover to prevent the escape of offensive vapours. One portion of phosphorus unites with the oxygen of the lime, and an acid is produced, which combines with some undecomposed lime. Another portion uniting with the calcium forms the phosphuret. When it is acted upon by water, it immediately inflames, phosphureted hydrogen being produced.—See 568.

868. CARBONATE OF LIME.—Symb. :C·Ca = carbonic acid 22 + lime 28.—Eq. 50. Solid, insoluble, heat expels the carbonic acid, muriatic acid and most other acids decompose it with effervescence. ICELAND SPAR, CALCAREOUS SPAR, MARBLE, CHALK, and LIMESTONE, are composed of this compound. Magnesia is often associated with it in some marbles and limestones in considerable quantity, besides smaller portions of other matters. Precipitated in a minute state of division, when an alkaline carbonate is added to a solution of a salt of lime. Carbonate of soda and muriate of lime = carbonate of lime and muriate of soda. Symb. :C·Na&HCl+·Ca = :C·Ca&HCl+·Na. The acid of the salt of lime used remains in solution with the alkali. The BICARBONATE is sparingly soluble in water.

869. MURIATE OF LIME.—Symb. HCl+·Ca· Prepared by saturating dilute muriatic acid with lime; carbonate of lime is generally employed. Symb. HCl&:C·Ca = HCl·Ca&:C. It is procured also as a residuum from many chemical operations. Crystallizes in prisms from a hot saturated solution. Converted into chloride of calcium when subjected to heat, the hydrogen of the muriatic acid combining with the oxygen of the lime.

Fig. 37.

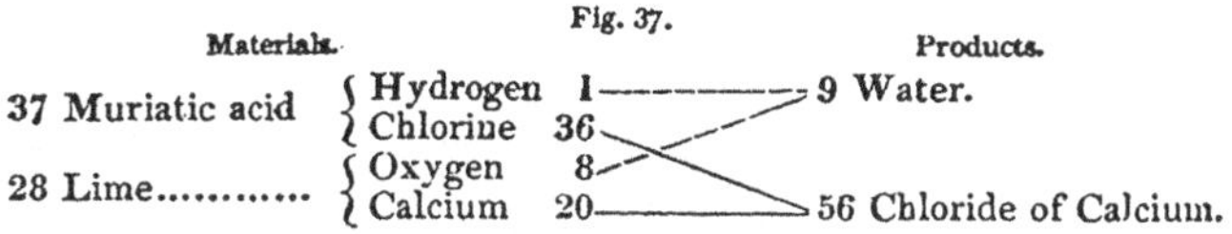

Symb. HCl+·Ca = ·H&Cl·Ca.

Decomposition of chloride of calcium by water. *Symb.* ·H&ClCa = HCl+·Ca.

870. Very deliquescent, and soluble in water. Liquefies snow rapidly, producing intense cold; its crystals contain 6 equivalents of water, and they act more powerfully than the chloride of calcium; by melting them, and then stirring them continually as the liquid cools, it is obtained in a minute state of division, extremely favourable for its action.

871. CHLORIDE OF LIME, commonly called BLEACHING POWDER.—White, with a slight shade of greenish-yellow colour; has an odour of chlorine, and a very strong harsh taste. Destroys vegetable and animal colouring matters, and various effluvia, in the same manner as chlorine. It contains variable quantities of chlorine, which is merely absorbed, according to some chemists, by the slaked lime employed in its preparation; but the more general opinion is, that the chlorine unites chemically with one portion of the lime, forming the chloride, with which an excess of lime is always mixed, according to the mode of preparation.

872. Muriatic and other acids unite with the lime and disengage chlorine with effervescence. Water dissolves it, a portion of lime being separated; it is much more soluble in water than lime. The quantity of chlorine associated with the

lime is generally estimated by finding its power of decolorizing a solution of indigo in sulphuric acid of a known strength.

873. By long keeping, and exposure to air, light, and moisture, muriatic and chloric acids are believed to be produced. When subjected to heat, part of the chlorine is evolved, and the rest unites with the calcium, disengaging oxygen.

CHAP. II. BARIUM.

874. Symb. Ba. Eq. 70.—Metallic, of a dark grey colour, heavier than water, attracts oxygen rapidly from air and water. Prepared in the same manner as calcium. See 846.

875. Baryta, or Oxide of Barium.—Symb. ·Ba = barium 70 + oxygen 8. —Eq. 78.—Name from βαρὺς, heavy; its compounds having a high specific gravity compared with other earthy minerals.

876. Baryta is similar to lime in all its leading properties. It is usually obtained in the form of a grey coloured mass, which slakes with water, and produces much heat. It is much more soluble in water than lime, 3 parts at 212°, and 20 at 60°, dissolving the hydrate, which consists of 20 eqs. water + 1 baryta. A saturated hot solution deposits crystals as it cools. Its solution is highly alkaline, turns blue colouring matter of cabbage and violets to a green, and turmeric to a brown. Precipitates carbonic acid from water and alkaline carbonates, but is soluble in an excess of carbonic acid. Sulphuric acid forms a very insoluble compound with baryta, and is much employed as a test of this substance in solution; the sulphate precipitated appears in the form of a white powder.

877. Baryta and all its compounds, except the insoluble sulphate, are poisonous. Its salts are usually obtained by the action of diluted acids with baryta, with the carbonate of baryta, or with the sulphuret of barium.

878. Baryta is generally procured by decomposing the nitrate in a crucible at a red heat; all the nitric acid is expelled.

879. It may also be obtained, as Dr Hope shewed, by subjecting the carbonate to an intense heat in a crucible. Carbonic acid has a much greater attraction for baryta than for lime, and hence the higher temperature required for its decomposition. Symb. :C·Ba = :C&·Ba.

880. Binoxide of Barium, or Peroxide of Barium.—Symb. :Ba = barium 70 + oxygen 16. Eq. 86.—Employed in preparing oxygenated water. Formed in the manner described in 847.

Compounds of Barium and Baryta.

881. Nitrate of Baryta.—Symb. ::N+·Ba = nitric acid 54 + baryta 70. Eq. 124.—Crystallizes in octohedrons, which contain no water of crystallization, decrepitate, fuse, and are ultimately decomposed by heat. Soluble in 11 parts of cold and in about 3 of boiling water.

882. Prepared by saturating diluted nitric acid with baryta, using the carbonate or hydrosulphuret; the latter is procured by the action of water upon the sulphuret of barium. Carbonic acid or sulphureted hydrogen are disengaged as the nitric acid is neutralized.

883. Much employed as a test of sulphuric acid.

884. Sulphate of Baryta, or Heavy Spar.—Symb. :S·Ba = sulphuric acid 40 + baryta 78. Eq. 118.—Solid, tasteless, insoluble, infusible.—Sp. gr. 4.4. Occurs native in tabular or prismatic crystals; always formed and precipitated as a white powder when solutions containing sulphuric acid are mixed with solutions containing baryta. Employed in preparing salts of baryta. When decomposed

by charcoal (713) it is reduced to SULPHURET OF BARIUM. Symb. ⁝S·Ba&4C= SBa&4·C. This substance decomposes water, producing sulphureted hydrogen and baryta, which remain in union, Symb. SBa&·H = HS+·Ba ; the sulphureted hydrogen is easily expelled by most other acids.

885. CARBONATE OF BARYTA, or WITHERITE.—Symb. :C·Ba = carbonic acid 22 + 78 baryta. Eq. 100.—Solid, insoluble, poisonous. Effervesces with acids; much employed, like the sulphuret of barium, in preparing salts of baryta.

886. CR. MURIATE OF BARYTA.—Symb. ·H+(HCl+·Ba).—By heat it becomes CHLORIDE OF BARIUM, Symb. ClBa ; the hydrogen of the acid uniting with the oxygen of the baryta, and forming water, which is expelled. Soluble in 5 parts of water at 60°. Crystallizes in 4-sided tables.

887. PREPARED by neutralizing baryta with muriatic acid : the carbonate may be used, Symb. HCl&:C·Ba = HCl·Ba&:C ; or the hydrosulphate (hydrosulphuret), HCl&HS·Ba = HCl·Ba&HS.

CHAP. III. STRONTIUM.

888. Symb. Sr.—Eq. 44.—It is similar to barium in all its leading characters.

889. STRONTIA, or OXIDE OF STRONTIUM.—Symb. ·Sr = strontium 44 + oxygen 8.—Eq. 52.—Its properties are similar to those of baryta (see 876). A saturated hot solution deposits 4-sided tabular crystals as it cools. These contain 12 eqs. water + 1 eq. of strontia ; they are soluble in 50 parts of water at 60°, and in 2 at 212°. By heat, they fuse and lose all their water of crystallization, excepting 1 eq. The salts of strontia communicate a deep red tint to inflammable matter in a state of combustion, by which they are easily distinguished from those of baryta. It is distinguished from lime by the insolubility of the sulphate, which is much less soluble than sulphate of lime, and generally said to be insoluble, though water can take up a minute quantity.

890. PREPARED most conveniently by heating the carbonate to whiteness in a crucible, when the carbonic acid is expelled.

891. MURIATE AND NITRATE OF STRONTIA are usually prepared by saturating muriatic and nitric acids with strontia, using for this purpose the carbonate of strontia, or the hydrosulphate (hydrosulphuret) of strontia, the latter being procured from the sulphuret of strontium by the decomposition of water. Sulphate of strontia decomposed by charcoal gives sulphuret of strontium. Sec 713.

CHAP. IV. MAGNESIUM.

892. Symb. Mg.—Eq. 12.—Resembles silver in appearance, fuses at a red heat, and burns then in air or oxygen, producing magnesia. Prepared by passing potassium in vapour over chloride of magnesium. The potassium unites with the chlorine, which leaves the magnesium; water removes the compound of chlorine and potassium.

893. MAGNESIA, or OXIDE OF MAGNESIUM.—Symb. ·Mg = magnesium 12 + oxygen 8.—Eq. 20.

894. Magnesia is white, inodorous, does not slake like lime, and requires upwards of 5000 parts of water for solution, according to Dr Fyfe. It is much less caustic than the preceding earths, has a feeble alkaline taste, and affects vegetable colouring matters in the same manner as the alkalis, when they are mixed with the magnesia in the solid form ; the solution in water scarcely affects them, the quantity dissolved being so trifling.

895. PREPARED usually from the carbonate by heat, which expels the carbonic acid.

896. Magnesia is distinguished by the great solubility of its sulphate, by its solutions not precipitating the common carbonate of ammonia, and by the decomposition of its soluble salts with lime-water, which separates the magnesia. Oxalate of ammonia precipitates lime and separates it from magnesia, which is not precipitated if the solution be dilute. Magnesia, quite free from carbonic acid, does not effervesce with acids.

897. NITRATE OF MAGNESIA.—Symb. ⁝:N+·Mg.—Soluble in water and alcohol; deliquescent.

898. CR. SULPHATE OF MAGNESIA.—Symb. 7·H+(⁝S·Mg) = water 63 + sulphate of magnesia 60.—Eq. 123.—Commonly called EPSOM SALTS. Soluble in an equal weight of cold water, and in 3–4ths of boiling water; crystallizes in 4-sided prisms; its water of crystallization is expelled by heat, the dry sulphate remaining. Decomposed by potassa, soda, and their carbonates.

899. PREPARED from BITTERN, the drainings from sea-water which has been evaporated for the preparation of common salt, being purified by repeated solution and crystallization. Prepared also by communicating sulphuric acid to the magnesia in magnesian limestone.

900. *Distinguished from oxalic acid* by its bitter taste, by its not affecting the vegetable blues, nor effervescing with alkaline carbonates, and also by not being dissipated by heat.

901. PHOSPHATE OF AMMONIA AND MAGNESIA.—Insoluble in water, dissolved by diluted nitric, muriatic, and acetic acids, decomposed by heat, the ammonia being expelled. Precipitated from solutions containing ammonia and magnesia on the addition of phosphoric acid, or a solution of a phosphate.

902. CARBONATE OF MAGNESIA.—Symb. :C·Mg = carbonic acid 22 + magnesia 20.—Eq. 42.—Solid, white, tasteless, inodorous, insoluble in water. Decomposed rapidly with effervescence by dilute sulphuric, nitric, and muriatic acids. Procured by decomposing a solution of sulphate of magnesia by a hot solution of carbonate of potassa or soda; the carbonate of magnesia is precipitated, and an alkaline sulphate remains in solution. Sulphate of magnesia 60 and carbonate of soda 54 = sulphate of soda 72 and carbonate of magnesia 42. Symb. ⁝S·Mg&:C·Na = ⁝S·Na&:C·Mg.

903. BICARBONATE OF MAGNESIA may be prepared in solution by passing a stream of carbonic acid gas through water containing magnesia suspended in it. When a bicarbonate of an alkali is added to a solution of a salt of magnesia in water, no precipitate is formed; but, by boiling, the excess of carbonic acid is expelled, and then carbonate of magnesia appears.

904. MURIATE OF MAGNESIA is procured by evaporating sea-water till the most of the other salts shall have been separated. It is very deliquescent.

II. COMMON TERRIGENOUS METALS.

905. The common terrigenous metals produce oxides which are insoluble in water, and have none of the alkaline reaction which the preceding earths present.

CHAP. I. ALUMINUM.

906. Symb. Al.—Eq. 10?—Metallic lustre, resembling platinum. Prepared from the chloride of aluminum by potassium, which combines with the chlorine and detaches the aluminum.

907. ALUMINA, or OXIDE OF ALUMINUM.—It consists of 1 eq. of oxygen and 1 of aluminum; but many chemists have considered it as a compound of 2 eqs. of aluminum and 3 of oxygen. White, tasteless, inodorous, insoluble in water, but can unite with it in various proportions, forming hydrates. It combines with acids and alkalis, and produces with the fixed alkalis compounds which are soluble in water. Alumina does not combine with carbonic acid.

908. Prepared from a solution of alum by adding potassa in solution, the alumina being precipitated; an excess of alkali must be avoided, as it would redissolve the alumina.

909. Alumina is extensively distributed throughout the mineral kingdom; it exists in all kinds of clay, and is found nearly pure in the sapphire and ruby.

910. ALUM, or SULPHATE OF ALUMINA AND POTASSA.—Symb. $3\dddot{S}\cdot Al+\dddot{S}\cdot K+25\cdot H$ = 3 eqs. of sulphate of alumina + 1 of sulphate of potassa + 25 of water. Crystallizes in octohedrons. Soluble in 5 of water at 60°, and in little more than an equal weight of water at 212°. Taste sweet, acid, and astringent. Reddens litmus powerfully, has little effect on the vegetable blue of cabbage. Decomposed by alkalis, alkaline carbonates, lime, and many other substances. Exposed to heat, it undergoes the watery fusion; when the water has been expelled, a spongy porous mass is left.

911. Exposed to heat with sugar, flour, or other carbonaceous substances, it is completely decomposed, and a compound remains, spontaneously inflammable: it is usually termed HOMBERG'S PYROPHORUS. Sulphurets of aluminum and potassium are formed during the decomposition. Burns brilliantly in oxygen gas, and with a dull red light when exposed to the air.

912. Alum of various kinds may be procured by substituting ammonia, soda, oxides of chrome, iron, or manganese, for potassa.

CHAP. II. SILICUM.

913. Symb. Si.—Eq. 8?—Colour dark brown, uninflammable, not oxidated by sulphuric or nitric acids, has no metallic lustre. Its oxide does not combine with the acids and form salts, and hence Dr Thomson and many others have regarded silicum as more analogous to carbon and boron than to the metals; they accordingly term it Silicon.

914. SILICA, SILICIC ACID.—Symb. $\dot{Si}$ = oxygen 8 + silicum 8.—Eq. 16.—Insipid, insoluble, inodorous, infusible, except by the oxyhydrogen blowpipe. Rough and hard to the touch. Exists in large quantities in numerous minerals. Rock-crystal, and some kinds of sand, flint, and other minerals, contain silica nearly pure.

915. COMPOUNDS OF SILICA.—Silica has little or no attraction for any of the acids excepting the fluoric. With the fixed alkalis and various metallic compounds, it forms glass and numerous other important vitreous combinations.

916. SILICATED HYDROFLUORIC ACID, called also FLUOSILICIC ACID.—Sp. gr. 3.611.—Gaseous, transparent, colourless, pungent and suffocating, produces white fumes with air containing moisture. Does not support combustion. Water condenses 365 times its volume, but 1–3 of the silica is deposited; it must be collected at the mercurial trough; corrodes glass only when moistened.

917. It is prepared by heating 1 part of fluor-spar with 1 of glass and 2 of sulphuric acid; the sulphuric acid decomposes the fluor-spar, producing hydrofluoric acid, in the manner described in 698. This acid was supposed till lately merely to combine with the silica; but the more general opinion now is, that the hydro-

gen of the acid unites with the oxygen of the silica forming water, the fluorine and silicum alone combining to form the gas. According to this view, the term fluosilic gas is to be preferred.

918. Silicated Potassa.—This compound may be prepared with various proportions of silica and potassa. To effect the combination, silica is generally fused with the carbonate of potassa, the carbonic acid being expelled during the application of the heat. With 3 of carbonate of potassa and 1 of silica, a glass is formed which is soluble in water and deliquescent; if an excess of silica be used with the alkaline matters, the glass procured is insoluble in water. Flint Soap is a compound which is likely to be very generally employed; it is merely a variety of silicated alkali, produced by boiling finely powdered flint with potassa or soda.

919. Common Bottle Glass consists of impure silica, as common sea-sand, fused with impure alkaline salts or other substances.

920. Window Glass is made with a purer sand and alkali. Soda, from kelp or barilla, is the alkali generally preferred.

921. Plate Glass is made with the purest alkali (soda) and siliceous matter.

922. Common Flint Glass contains about 50 of silica with 34 of the oxide of lead and 14 of potassa. The oxide of lead gives density, lustre, and increased fusibility to the glass. About 1-1200 part of the peroxide of manganese removes the green colour which a small quantity of oxide of iron contained in the material used is apt to communicate: it gives oxygen to the oxide of iron.

923. Solutions of silicated potassa are decomposed by nitric, sulphuric, and other acids, which neutralize the potassa, and precipitate the silica. When it meets with the water in its nascent state, this fluid is apt to dissolve it, but by evaporating to dryness, the silica becomes hard, and quite insoluble in water.

924. Muriate of ammonia is often used to precipitate silica instead of the uncombined acids; the ammonia is left free, when the acid unites with the alkali previously combined with the silica.

925. Five parts of carbonate of potassa with four of carbonate of soda, form a very fusible alkaline mixture, much employed in examining siliceous compounds.

CHAP. III. GLUCINUM, ITTRIUM, ZIRCONIUM, THORINIUM.

926. These metals are extremely rare compared with the other terrigenous metals.

927. Glucinum with oxygen forms the earth Glucina, found in the beryl and in the emerald.

928. Ittrium with oxygen forms the earth Ittria, which exists in the mineral called Gadolinite.

929. Zirconium and oxygen form Zirconia, one of the ingredients of the Zircon of Ceylon.

930. Thorinum and oxygen form Thorina, an earth discovered by Berzelius in a Norwegian mineral.

ORDER III. CALCIGENOUS METALS.*

I. COMMON METALS WHOSE OXIDES CANNOT BE REDUCED BY HEAT ALONE.

CHAP. I. IRON.

931. Symb. Fe.—Eq. 28.—Sp. gr. 7.7.—Fused only by the greatest heat of a wind or blast furnace. Colour gray, lustre brilliant when polished, malleable and ductile, but not so much so as gold, silver, and some other metals. In tenacity it exceeds all metals. When heated, it can be worked into any shape, or joined to another piece by *welding* or hammering them together. It is attracted by the magnet, and can be rendered permanently magnetic.

932. DISTRIBUTION. Extensively distributed throughout the mineral kingdom. Small detached masses have been found nearly pure, and probably of meteoric origin. The principal ores are the OXIDES of iron, which afford the purest metal, and the CLAY IRON-ORE, which consists of oxide of iron united with carbonic acid, alumina, silica, and other earthy matters.

933. PREPARATION OF IRON.—The clay iron-ore, after being roasted, is heated with charcoal, coke, or coal, and lime, in a blast-furnace. The carbon removes the oxygen from the iron, forming carbonic oxide or carbonic acid, which escapes. The lime combines with the earthy matter of the ore, forming thereby a fusible compound, through which the melted iron falls with facility, and collects at the bottom of the furnace. The SLAG, or compound of the various earthy matters, is run off from time to time, and the iron is allowed to flow into moulds made in sand, where it soon consolidates, and forms what is termed PIG IRON. The oxide is heated with charcoal alone.

934. In preparing this metal the hot-blast introduced by Mr Neilson and Mr Dunlop in the Clyde Iron-Works, has enabled them to prepare the metal with much less fuel and lime than were formerly necessary. See 579.

935. The presence of charcoal, and small portions of the metals aluminum and silicum, render the metal procured in this manner more brittle and fusible than pure iron: it is usually called CAST-IRON; this fuses at 2786° according to Daniell. Two varieties have been more particularly noticed,—

936. 1*st*, WHITE CAST-IRON. It is very hard and brittle.

937. 2*d*, GRAY CAST-IRON: which is softer and less brittle.

938. Malleable-iron is prepared from cast-iron by heating it in a furnace, burning off carbonaceous matter by a strong blast made to play on the surface, and removing the earthy impurities, which rise to the top. As its purity and tenacity increase, its fusibility diminishes, and, after long continued hammering and pressure between rollers at an elevated temperature, it forms the malleable iron of commerce; called also *Forged* Iron, *Wrought* Iron, *Soft* Iron.

939. OXIDES OF IRON.—Iron is oxidated by air and water, by a number of acids and other substances. But air perfectly dry and water perfectly free from air and acid do not oxidate iron.

940. OXIDE OF IRON.—Symb. $\dot{Fe}$ = iron 28 + oxygen 8.—Eq. 36.—Colour dark, seldom obtained pure, generally associated with a little peroxide; separated

* See 707 and 737.

from its salts in solution as a white HYDRATE by alkalis. Formed when iron burns in air or oxygen gas. At a white heat iron burns rapidly, even when cold air is blown upon it, the oxide produced being rapidly melted. Watery vapour passed over iron-turnings at a red heat produces the protoxide, hydrogen being evolved. And if the hydrogen be transmitted over the oxide at a similar temperature, it combines with the oxygen, the metal being restored; if the oxide be in a minute state of division, and reduced at a temperature below that of a red heat, the metal is left in a minute state of division, and inflames on contact with the air. Exposed to air and moisture, the oxide soon becomes peroxide. Its salts are generally of a green colour.

941. PEROXIDE OF IRON.—Symb. $\vdots Fe^2$ = iron 56 + oxygen 24.—Eq. 80.—Colour deep red, not attracted by the magnet. Prepared from sulphate of iron by heating it in a crucible. All the acid is expelled, and 2 eqs. of oxide taking 1 eq. of oxygen from 1 eq. of sulphuric acid, become 1 eq. of peroxide, the sulphurous acid from the decomposed sulphuric acid being disengaged along with some sulphuric acid and water not decomposed.

Fig. 38.

Materials.			Products.
Sulphate of Iron 2 eqs.	Sulphuric Acid	40	40 Sulphuric Acid.
	Sulphurous Acid	32	32 Sulphurous Acid.
	Oxygen	8	
	Oxide of Iron ...	36	
	Oxide of Iron ...	36	80 Peroxide of Iron.

Symb. $2\vdots S{\cdot}Fe = \vdots Fe^2 \& \vdots S \& :S$.

942. IRON RUST consists principally of this oxide in union with water. Carbonic acid and ammonia are also found in small quantities, the acid being derived from the air, and the ammonia formed by the combination of the nitrogen of the air, with the hydrogen of the same decomposed water.

943. The BLACK OXIDE OF IRON, regarded by many as a distinct oxide, is generally considered a mixture of the oxide and peroxide in various proportions.

944. IRON AND NITRIC ACID.—Iron decomposes nitric acid rapidly, being oxidated, and combining with a portion of undecomposed acid; if to a solution of nitrate of iron, prepared in the cold with an excess of acid, carbonate of potassa be added in excess, nitrate of potassa is formed, the oxidated metal is precipitated, and immediately redissolved in the excess of carbonate of potassa, forming a very deep coloured liquid. This solution has been used medicinally under the name of LIQUOR FERRI ALKALINI.

Nitric acid is often used to convert protoxide into peroxide of iron.

945. SULPHUR AND IRON, Sulphuret of Iron.—Symb. SFe = iron 28 + sulphur 16.—Eq. 44.—Brittle, of a dark grey colour, much employed in the preparation of sulphureted hydrogen. Procured most conveniently for this purpose by bringing sulphur in contact with iron at a red heat; the compound immediately falls down in the fluid form, and may be collected in water. If 28 parts of iron filings be mixed with 16 of sulphur, and the mixture heated in a Florence flask, they unite, a glow of light attending the combination.

946. BISULPHURET OF IRON, or IRON PYRITES.—Symb. S^2Fe = iron 28 + 32 sulphur.—Eq. 60.—Hard, yellow, and cubical crystals; not acted upon by any of the common acids except the nitric. Occurs in large quantity in nature; it is often heated for the preparation of sulphur, part of this element being expelled from it in this manner.

947. If 1 part of sulphur with 2 of iron-filings be moistened and exposed to the air, they absorb oxygen and become warm, frequently taking fire; many native

sulphurets of iron produce the same effect with air and moisture. Sulphate of iron is formed after long exposure.

948. SULPHATE OF IRON, called also Copperas or Green Vitriol.—Symb. 7·H + (⁝S·Fe) = water 63 + sulphuric acid 40 + oxide of iron 36.—Eq. 139.—Melted by heat, most of the water of crystallization being soon evaporated, and a greenish-white sulphate left; by a higher temperature it is resolved into peroxide of iron, see 941. Soluble in about 2 of cold water, and in less than its weight of boiling water; crystallizes in rhomboidal prisms of a bluish-green at first, which becomes green and ultimately red as they absorb oxygen from the air. Taste metallic and stiptic.

949. Its solution with water becomes turbid on exposure to the air, oxygen being absorbed and a terpersulphate left in solution, while a dipersulphate is precipitated.

950. Used in the preparation of ink, peroxide of iron, carbonate of iron, prussian blue, dyeing black, &c.

951. PREPARED by the action of sulphuric acid, water, and iron (see 435). Formed also by the oxygenation of native sulphuret of iron exposed to the action of air and moisture.

952. TERPERSULPHATE OF IRON.—Symb. 3⁝S+⁝Fe^2 = 3 sulphuric acid 120 + peroxide of iron 80.—Eq. 200.—Formed by heating sulphate of iron with sulphuric and nitric acids. Colour red, soluble, not crystallizable.

953. DIPERSULPHATE OF IRON.—Symb. ⁝S+2⁝Fe^2 = sulphuric acid 40 + peroxide 160.—Eq. 200.—Precipitated from sulphate of iron in solution when exposed to the air.

954. *Compounds with Carbon.*—CARBURET OF IRON or STEEL is composed of pure iron, combined with a very small quantity of carbon; it is procured by heating malleable iron imbedded in powdered charcoal, and is particularly distinguished by the hardness which it acquires when heated to redness and suddenly cooled. The different degrees of temperature to which it is heated before it is cooled, give rise to those varieties in its state of cohesion, or *temper*, as it is technically called, by which it is so extremely well adapted for different instruments.

955. PLUMBAGO, BLACK LEAD, or GRAPHITE, is the name given to a compound consisting almost entirely of charcoal with a small quantity of iron, varying from 4 to 10 per cent. The native compound is much used in making pencils.

956. CARBONATE OF IRON.—Symb. :C·Fe = carbonic acid 22 + oxide of iron 36.—Eq. 58.—Has a greenish-white colour when first precipitated, but soon darkens, and ultimately much of it loses carbonic acid, attracting oxygen when exposed to the air, peroxide of iron being formed. In this condition it has long been called SUBCARBONATE OF IRON, but, as Mr Phillips shewed, it consists principally of peroxide of iron, with a small and variable proportion of the carbonate. It is similar in composition to RUST OF IRON. No percarbonate (carbonate of the peroxide) has been discovered.

957. Iron is frequently met with in mineral waters, where it is retained in solution in the form of a supercarbonate. By boiling, the excess of carbonic acid is expelled, and an ochry sediment deposited.

958. Carbonate of iron is prepared from sulphate of iron, 76 parts, mixed in solution with 54 of carbonate of soda; 58 of carbonate of iron are precipitated, and 72 of sulphate of soda remain in solution.—Symb. ⁝S·Fe&:C·Na = ⁝S·Na&:C·Fe. By washing it with water, and preventing it from being exposed to the air, it may be retained in the form of a carbonate. Dried, in contact with air, it is always decomposed.

959. Acetate of Iron is procured by digesting carbonate or oxide of iron, with diluted acetic acid. Acetate of potassa and sulphate of iron, when well mixed with alcohol, give a tincture of acetate of iron, and sulphate of potassa which is left undissolved.

960. Tartrate of Potassa and Iron is formed when bitartrate of potassa is mixed with iron filings, moistened, and exposed to the action of the air; the iron, as it is oxidated, uniting with the excess of acid in the bitartrate. It has a less offensive metallic taste than any of the other preparations of iron.

961. Chloride of Iron is procured when the muriate of iron is decomposed cautiously by heat, the hydrogen of the acid uniting with the oxygen of the oxide. The Muriate of Iron is formed when iron is put into aqueous muriatic acid, previously diluted with an equal bulk of water; the iron is oxidated by the water, hydrogen gas is evolved, and the acid combines with the oxide of iron.

962. Perchloride of Iron is formed when iron is burned in chlorine gas. By water it is converted into Permuriate of Iron, part of the water being decomposed, while the rest dissolves the new compound. Permuriate of iron is generally prepared by dissolving peroxide of iron in muriatic acid. It has a reddish colour, and is soluble in alcohol. The muriate is not soluble in alcohol. The permuriate is usually made with an excess of acid, but the oxide and acid can unite in many different proportions. Mr Phillips has found that 1 equivalent of muriatic acid can combine even with 10 of peroxide of iron.

963. Muriate of Ammonia and Iron is formed when muriate of ammonia and muriate of iron are heated together and sublimed.

964. Hydriodate of Iron is formed when water, iodine, and iron, are mixed together and heated. The iron takes oxygen, and the iodine takes hydrogen from part of the water. The hydriodic acid and oxide of iron thus formed combine and remain in solution; the liquid is transparent and colourless, but soon acquires a deep tint on exposure to the air.

Tests of Iron.

965. Gallic Acid precipitates the dark coloured gallate of iron, the colouring matter of ink, when added to persalts of iron. It has no effect on the protoxide. Acids, alkalis, and chlorine, decompose the colouring matter of ink. It may often, however, be made to reappear again by the action of an excess of gallic acid, or the form of letters effaced on paper may be restored by hydrosulphate of ammonia, ferroprussiate of potassa, and sulphocyanate of potassa.

966. Ferroprussiate of Potassa gives a white precipitate with the oxide, and a deep blue with the peroxide; when the solution contains both oxide and peroxide, the precipitate presents a variety of tints, from a white with a light shade of blue to a deep blue, according to the quantity of peroxide present.

967. Sulphocyanate of Potassa gives a very brilliant red colour to solutions containing peroxide of iron; it does not affect solutions containing the oxide.

968. Hydrosulphuric Acid (sulphureted hydrogen) does not affect solutions containing oxide of iron; a milk-white precipitate is formed with solutions containing peroxide. Hydrosulphuret of ammonia gives precipitates of a dark colour with the salts of the oxide and peroxide.

969. Benzoate and Succinate of Ammonia are much employed to separate peroxide of iron, forming very insoluble compounds with it.

970. Alkalis and Alkaline Earths precipitate the oxides of iron from their solutions as hydrated oxides.

971. Alkaline Carbonates precipitate carbonates of the oxide; but when

they are added to solutions containing the peroxide, it is precipitated without any carbonic acid, for which it has little or no attraction.

972. Nitric Acid is often employed, in experiments with iron, to convert the oxide into peroxide; the iron is then more easily detected. A few drops of nitric acid, boiled with a small quantity of any solution containing iron, are sufficient for this purpose.

CHAP. II. LEAD.

973. Symb. Pb.—Sp. gr. 11.55. Melts about 600°. Colour bluish-gray, brilliant lustre, tarnishes quickly in the air. Malleable, ductile, sectile. Eq. 104.

974. Prepared from the sulphuret of lead or Galena, by heating it with coal and lime; the lime assists in removing the sulphur, and any sulphurous or sulphuric acid formed during the process. Iron may be used on the small scale.—Sulphuret of Lead and Iron = Sulphuret of Iron and Lead. Symb. SPb&Fe = SFe&Pb.

975. Precipitated slowly in crystals, forming the lead tree, when zinc is introduced into any of its solutions. The zinc takes all the acid and oxygen that may have been associated with the lead, replacing it in the solution.

Fig. 39.

Materials.			Products.
34 Zinc		34	93 Acetate of Zinc.
163 Acetate of Lead	Acetic Acid	51	
	Oxide of Lead: Oxygen	8	
	Oxide of Lead: Lead	104	104 Lead.

Symb. *A*·Pb&*Z* = *A*·*Z*&Pb.

976. *Oxides of Lead.*—Air and water charged with carbonic and other acids, oxidate lead. If the acid be the phosphoric, arsenic, sulphuric, or hydriodic, an insoluble crust is formed on the surface of the lead, which protects it from farther action, and hence water containing even a minute portion of these acids or their salts, may be kept with safety in leaden cisterns. But where the water is charged with carbonic acid, the carbonate not forming any protecting crust, mixes with the water in a minute state of division, and renders it poisonous.—See Christison on Poisons.

977. Nitric acid is used for oxidating and dissolving lead.

978. Dinoxide of Lead.—Symb. ·Pb^2 = lead 208 + oxygen 8.—Eq. 216.—Found when the oxalate of lead is heated to dull redness, the air being excluded.

979. Oxide. Called also Massicot, or Yellow Lead.—Symb. ·Pb = lead 104 + oxygen 8.—Eq. 112.—Insoluble, fusible at a red heat, and combines with numerous earthy and saline substances, forming very fusible compounds. Decomposed when heated with charcoal and other inflammable matters, metallic lead being obtained.

980. Prepared by heating metallic lead in contact with air. When imperfectly fused, so as to form small scales, which are mixed with a little red oxide of lead, it is called Litharge, or Semivitrified Oxide of Lead. Precipitated from solutions of its salts in combination with water, and of a white colour, on the addition of alkalis. This is the only oxide which exists in the salts of lead.

981. Peroxide of Lead.—Symb. :Pb = lead 104 + oxygen 16.—Eq. 120.

982. Prepared from red oxide of lead by nitric acid diluted with water, which dissolves and removes the oxide, leaving the peroxide.

983. Red Oxide of Lead, called also Deutoxide, or Minium.—Composed of variable proportions of oxide and peroxide (Phillips).

984. Prepared by exposing lead to air at a temperature about 700°.

985. NITRATE OF LEAD.—Symb. ::N·Pb = oxide of lead 112 + nitric acid 54. —Eq. 166.—Soluble in water, crystallizes in octohedrons.

986. SULPHURET OF LEAD.—Symb. SPb = sulphur 16 + lead 104.—Eq. 120. —The GALENA of mineralogists. The most abundant ore of lead; formed artificially when sulphur and lead are melted together, or when sulphureted hydrogen decomposes oxide of lead.—See 711.

987. SULPHATE OF LEAD.—Symb. :S·Pb = sulphuric acid 40 + oxide of lead 112.—Eq. 152.—White, very insoluble, precipitated when sulphuric acid and oxide of lead meet in solution. Symb. ::N·Pb&:S·Mg = ::N·Mg&:S·Pb.

988. PHOSPHATE OF LEAD.—Very insoluble, precipitated when phosphoric acid and oxide of lead meet in solution.

989. CARBONATE OF LEAD, or CERUSSE.—Symb. :C·Pb = carbonic acid 22 + oxide of lead 112.—Eq. 134.—Insoluble in water, very heavy, precipitated when an alkaline carbonate is added to a solution of a salt of lead. Symb. ::N·Pb& :C·Na = ::N·Na&:C·Pb.

990. CR. ACETATE OF LEAD. SUGAR OF LEAD.—Crystallizes in 6-sided prisms, soluble in water, taste sweet and astringent, gives a slight precipitate with carbonic acid, which is dissolved by a small excess of acid. Contains 1 eq. of acid, 1 eq. of oxide, and 3 eqs. of water.

991. PREPARED by neutralizing acetic acid with oxide of lead.

992. SUBACETATE OF LEAD, or GOULARD'S EXTRACT.—More easily decomposed than the acetate. Prepared by boiling oxide of lead in excess in dilute acetic acid; it is obtained with different proportions of oxides, a ditacetate or tritacetate being procured; the former contains 2 and the latter 3 eqs. of oxide, united with 1 of acid.

993. TARTRATE OF LEAD.—White, insoluble in water, dissolved by diluted nitric acid.

994. CHLORIDE OF LEAD.—Symb. ClPb = chlorine 36 + lead 104.—Eq. 140. —Sparingly soluble in water, deposited in crystals from a hot concentrated solution.

995. PREPARED when muriatic acid or its salts meet oxide of lead in solution; the hydrogen of the acid unites with the oxygen of the oxide, forming water, and the chlorine combines with the metal.

996. IODIDE OF LEAD.—Symb. IPb = iodine 124 + lead 104.—Eq. 228.—Colour brilliant yellow, crystalline, formed when hydriodic acid meets with oxide of lead in solution; the oxide decomposes the hydriodic acid, the hydrogen of the acid and the oxygen of the oxide forming water, while the metal and the iodine combine.

Fig. 40.

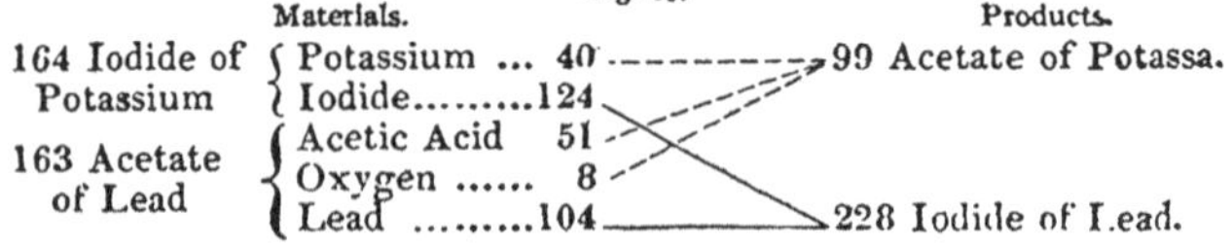

Symb. *A*·Pb&IK = *A*·K&IPb.

Tests of Lead.

997. SULPHURETED HYDROGEN and HYDROSULPHATE OF AMMONIA give a dark brown or black precipitate.

998. ZINC precipitates metallic lead.

999. HYDRIODIC ACID gives a yellow precipitate of iodide of lead. Sulphuric acid precipitates white sulphate of lead.

CHAP. III. COPPER.

1000. Symb. Cu.—Eq. 32.—Colour reddish. Sp. gr. 8.8. Malleable, ductile, and sonorous; fuses at 1996°. Little affected by dry air, but becomes covered with a green crust of carbonate of copper in a moist atmosphere.

1001. Prepared principally from the native sulphuret of copper and iron, which is heated first by itself, and then with charcoal and siliceous fluxes. In the first part of the process, part of the sulphur is expelled and the metals oxidated; the oxidated iron forms a slag with the flux, and when the remainder of the sulphur has been expelled, the charcoal reduces the oxide of copper.

1002. Metallic copper is often prepared in a very pure form, by introducing iron into solutions of its salts, the iron uniting with all the substances previously combined with the copper in solution, and precipitating it in the metallic form.

1003. Dinoxide of Copper, formerly called Protoxide.—Symb. $\cdot Cu^2$.—Colour reddish. Prepared by heating the oxide with metallic copper, or by separating it from the salts of the dinoxide by alkalis. It is soluble in a solution of ammonia, producing a colourless solution, which attracts oxygen quickly from the air, and becomes blue.

1004. Oxide of Copper.—Symb. $\cdot Cu$ = copper 32 + oxygen 8.—Eq. 40.—Black, insoluble in water; with acids it forms coloured salts. (Other oxides have been described, but they are of less importance.)

1005. Prepared by heating nitrate of copper, all the nitric acid being expelled; or by adding potassa in excess to sulphate or other salts of copper, when a blue hydrated oxide is separated, which becomes black by heat. If a small quantity of potassa be added, disulphate of copper, of a greenish-blue colour, is precipitated.

1006. Nitrate of Copper.—Symb. ::N·C = nitric acid 54 + oxide of copper 40.—Eq. 94.—Very soluble, deliquescent, escharotic, crystallizes in blue coloured prisms, which become green on exposure to heat, losing part of their acid and water of crystallization.

1007. Prepared by the action of diluted nitric acid on metallic copper.—See Nitric Oxide.

1008. Sulphate of Copper, or Blue Vitriol, is soluble in 4 of cold and 2 of hot water; crystallizable, styptic, escharotic.

1009. Prepared on oxygenating the sulphuret by heat and air, or by dissolving the oxide in dilute sulphuric acid.

1010. Sulphuret of Copper.—Symb. SCu.—Formed in general artificially, when required pure, by heating a mixture of sulphur and copper filings; heat and light are evolved during the combination.

1011. Ammoniuret of Copper, may be formed by the action of ammonia in solution on the oxide of copper; it has a very deep blue colour. The pharmaceutic preparation, commonly called Ammoniuret of Copper, or Ammoniated Copper, contains sulphuric acid, oxide of copper and ammonia. It is procured by triturating in a mortar 2 parts of sulphate of copper and 3 parts of the common carbonate of ammonia. Carbonic acid escapes with effervescence, the mixture slowly becoming moist as the water in the salt of copper is liberated; part of the ammonia combines with the sulphuric acid, and part with the oxide. It has an intense blue colour, and becomes green when freely exposed to the air, or slightly heated, part of the ammonia escaping.

1012. Sulphate of copper, free from water, absorbs ammoniacal gas readily, and becomes of a blue colour. Other sulphates possess a similar property; the ammonia is easily expelled by heat.

1013. DICARBONATE OF COPPER.—Symb. :C+2·Cu = carbonic acid 22 + oxide of copper 80.—Eq. 102.—Of a green colour, insoluble; found in combination with water in the mineral called MALACHITE. Prepared artificially by adding carbonate of soda or potassa to sulphate of copper in solution. By boiling in water it becomes black; the dicarbonate loses its combined water and acid, the black oxide being left.

1014. ACETATE OF COPPER.—Soluble, crystallizes in deep green-coloured octohedrons, consisting of 40 oxide, 51 acetic acid, combined with water. Prepared by boiling dicarbonate or ditacetate of copper in diluted acetic acid.

1015. DITACETATE OF COPPER, containing 1 eq. of acetic acid, and 2 of the oxide, and TRITACETATE, composed of 1 eq. of acid and 3 of oxide, form two of the most important subacetates of copper. VERDIGRIS contains these salts in various proportions, and is prepared by corroding plates of copper with impure acetic acid.

1016. CHLORIDE and DICHLORIDE OF COPPER are formed when this metal, in a minute state of division, is introduced into chlorine. Muriatic acid and dicarbonate of copper give MURIATE OF COPPER of a fine green colour, soluble in water and alcohol, and resolved by heat into chloride of copper and water.

Tests of Copper.

1017. IRON and ZINC precipitate metallic copper.

1018. AMMONIA in excess gives a deep blue-coloured ammoniaret.

1019. FERROPRUSSIATE OF POTASSA gives a very peculiar mahogany precipitate.

1020. HYDROSULPHURET OF AMMONIA gives a very dark coloured precipitate.

1021. POTASSA and SODA precipitate the blue hydrated oxide.

CHAP. IV. ZINC OR SPELTER.

1022. Symb. Z.—Eq. 34.—Sp. gr. 7.—Colour bluish-white, texture crystalline. Brittle at ordinary temperatures unless it has been previously annealed. Malleable and ductile at 300°; melts at 700°; converted into vapour at a white heat.

1023. PREPARED from the carbonate of zinc by heating it with charcoal, the carbonic acid is expelled, the charcoal removes the oxygen of the oxide, the zinc is converted into vapour and condenses in the fluid form, after which it is allowed to drop into water. Zinc is not much affected by air and moisture at ordinary temperatures; but when any acid matter is present, it is quickly oxidated and often dissolved.

1024. OXIDE OF ZINC.—Symb. ·Z = zinc 34 + oxygen 8.—Eq. 42.—White when cold, yellow when hot, insoluble in water, forms soluble compounds with many of the acids, and is dissolved by solutions of the alkalis, forming transparent and colourless compounds.

1025. PREPARED by heating zinc in contact with the air, the metal burning with a bluish-green coloured flame; or on precipitating it from solutions of its salts by the alkalis, taking care to avoid any excess.

1026. NITRATE OF ZINC.—When nitric acid diluted with water is poured upon zinc, part of the acid and of the water are decomposed, and a little ammonia produced by the union of nitrogen from the acid and hydrogen from the water, as the metal is oxidated. The undecomposed acid unites with the oxide and the ammonia. Zinc acts with great rapidity on nitric acid.

1027. Sulphuret of Zinc, the zinc blende of mineralogists, is found native of a dark colour; precipitated from its solutions white and in combination with water, by hydrosulphuric acid. See 711.

1028. Sulphate of Zinc.—Symb. ⁝S·*Z* = sulphuric acid 40 + oxide of zinc 42.—Eq. 82.—Soluble, crystallizes easily in prisms from a strong hot solution; formed by oxygenating sulphuret of zinc as it is heated in contact with air, or by the action of sulphuric acid on zinc and water. See the preparation of Hydrogen.

1029. Carbonate of Zinc.—Symb. :C·*Z* = carbonic acid 22 + oxide of zinc 42.—Eq. 64.—Found native, forming the mineral called Calamine.

1030. Acetate of Zinc.—Soluble, crystallizable. Formed by the action of acetic acid on the oxide or carbonate of zinc, by the action of zinc on acetate of lead in solution, all the lead being precipitated and the zinc replacing it in the liquid, or by decomposing sulphate of zinc, as seen in the diagram.

Fig. 41.

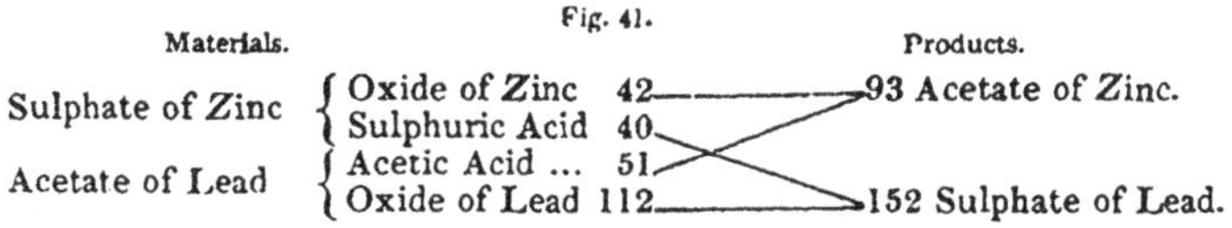

Symb. ⁝S·*Z*&:*A*·Pb = ⁝S·Pb&*A*·*Z*.

1031. Chloride of Zinc.—Formed when zinc is introduced into chlorine gas; or when dilute muriatic acid acts with metallic zinc, hydrogen being evolved, the solution being afterwards evaporated to dryness.

1032. Iodide of Zinc is prepared by heating iodine with zinc filings in water, and evaporating the solution to dryness.

Tests of Zinc.

1033. Hydrosulphate of Ammonia gives a white precipitate.

1034. Alkalis precipitate oxide of zinc, dissolving the oxide if they are added in excess.

CHAP. V. ANTIMONY.

1035. Symb. Sb; from *Stibium*, the old name of its most important ore; called also Regulus of Antimony.—Eq. 44.—Sp. gr. 6.8.—White, with a shade of blue, very brittle, lamellar texture, fuses at a red heat, burns in contact with air at an elevated temperature, producing much white smoke. Is not affected by a dry atmosphere, tarnishes slowly when exposed for a long time to the air.

1036. Prepared from the sulphuret by heating it in a crucible with iron filings; sulphuret of iron is formed, and the metallic antimony is separated.

1037. Oxide of Antimony.—Symb. ·Sb = antimony 44 + oxygen 8.—Eq. 52.—Prepared by several different processes:—

1038. I. By pouring muriate of antimony into a large quantity of water, it is resolved into two salts, a muriate with great excess of acid which remains in solution, and a submuriate composed of 9 eqs. of oxide and 1 of acid according to Phillips, which is precipitated. The latter has generally been called Oxide of Antimony, and has long been known by the name of Powder of Algaroth. All the acid may be removed from it by potassa in solution.

1039. It has a dull white colour, fuses by heat, and is volatilized by an increase of temperature. Exposed to heat and air when rough and porous, it absorbs oxygen and takes fire.

1040. II. By deflagrating equal weights of nitrate of potassa and sulphuret of antimony. The sulphur and antimony receive oxygen from the nitrate, and on removing the adhering potassa by water, the oxide appears of a dark brown colour, from being mixed with some sulphuret of antimony. It is, in this condition, the Crocus of Antimony of the Edinburgh College.

1041. III. By heating sulphuret of antimony slowly in the air, when most of the sulphur is burned away and the antimony oxidated. By fusing this oxide in a crucible it forms a vitreous mass when cold, and is called GLASS OF ANTIMONY. A little undecomposed sulphuret communicates a deep brown tint.

1042. IV. By boiling metallic antimony with aqueous sulphuric acid, it takes oxygen from part of the acid, disengaging sulphurous acid, and combining with the remainder. Water added in large quantity removes most of the acid, and leaves a subsulphate, which may be advantageously employed according to Phillips in the same manner as the preceding oxides.

Fig. 42.

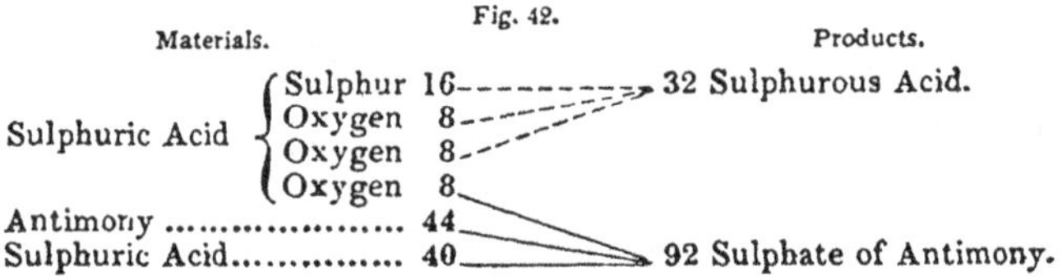

Symb. 2⋮S&Sb = ⋮S·Sb&:S.

1043. DEUTOXIDE OF ANTIMONY, or ANTIMONIOUS ACID.—Formed by exposing the oxide to heat and air, or by expelling by a red heat a portion of oxygen from the peroxide. White, insoluble in acids, combines with alkalis, not melted nor volatilized by heat.

1044. PEROXIDE OF ANTIMONY, or ANTIMONIC ACID.—Prepared, in combination with water, by heating metallic antimony with nitric acid to dryness, and then washing it with water. At 600°, the water is expelled and it becomes yellow.

1045. OXIDE OF ANTIMONY WITH PHOSPHATE OF LIME, or ANTIMONIAL POWDER.—Usually regarded as a mixture of peroxide (deutoxide ?) of antimony and phosphate of lime. Prepared by heating sulphuret of antimony mixed with hartshorn shavings in contact with air, the sulphur and all the animal matter in the hartshorn shavings being burned away, while the antimony is oxidated and left with the phosphate of lime. The proportion of oxide of antimony found in it varies, which accounts for the uncertainty of its action.

1046. SULPHURET OF ANTIMONY.—Symb. SSb = sulphur 16 + antimony 44. Eq. 60.—Colour bluish-grey, texture crystalline and striated. Precipitated of a bright orange colour by hydrosulphuric acid from solutions of antimony, but becomes of the usual colour when the water mixed with it is expelled by heat.

1047. Boiled in a solution of potassa, it is dissolved, and a precipitate appears of a dark brownish colour as the liquid cools, which is called KERMES MINERAL, consisting of sulphuret of antimony associated with water, and generally with variable proportions of oxide of antimony and a portion of potassa. Dilute sulphuric acid added to the remaining fluid precipitates GOLDEN SULPHURET OF ANTIMONY, which contains more sulphur than Kermes mineral and no potassa. If the dilute sulphuric acid be added to the solution in potassa while hot, the compound usually termed PRECIPITATED SULPHURET OF ANTIMONY is formed.

1048. TARTRATE OF ANTIMONY AND POTASSA, or TARTAR EMETIC.—Eq. 363 = 1 eq. of bitartrate of potassa 180 + 3 eqs. of oxide of antimony 156 (52×3)

+ 3 eqs. of water 27. Prepared by boiling the bitartrate and oxide of antimony in water, afterwards filtering the solution and crystallizing. It generally presents crystals which are transparent and colourless, but become opaque by keeping; its primary form is an octohedron with a rhombic base. Soluble in 3 parts of boiling and in 15 of cold water; its solution reddens litmus feebly, and is decomposed when long kept. Decomposed by heat, potassa and metallic antimony being left, when the air is admitted, and a pyrophorus formed when it is excluded. Decomposed also by acids, alkalis, alkaline carbonates, infusion of galls or of Peruvian bark, sulphureted hydrogen, and hydrosulphurets. Astringent vegetable solutions generally precipitate a compound of tannin and oxide of antimony. A solution of sulphate of magnesia does not decompose tartar emetic.

1049. Chloride, or Butter of Antimony.—Symb. ClSb = chlorine 36 + animony 44.—Eq. 80. — Formed with combustion when antimony filings are thrown into chlorine, or by subjecting them to heat with bichloride of mercury, when it is separated by distillation. It has the consistence of butter at common temperatures.

1050. Muriate of Antimony is formed when sulphuret of antimony is heated in common muriatic acid. Part of the water and the sulphuret being decomposed, oxide of antimony is formed, which combines with the muriatic acid, and sulphureted hydrogen gas escapes.

Fig. 43.

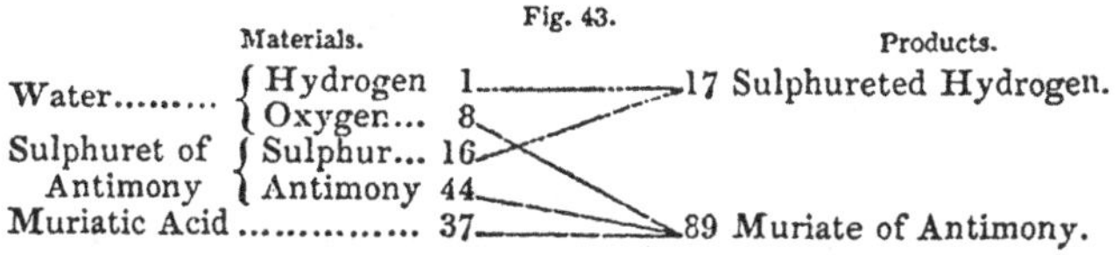

Symb. SSb&·H&HCl = HCl+·Sb&HS.

1051. *Tests of Antimony.*—This metal is particularly distinguished by the bright orange coloured precipitate which solutions of its oxide give with sulphureted hydrogen. Dr Turner recommends the precipitate to be collected and decomposed by hydrogen gas passed over it in a tube, where it is heated by a spirit-lamp; the sulphur being removed by the hydrogen, the metal is procured in a pure form.

CHAP. VI. TIN.

1052. Symb. Sn.—Eq. 58.—Sp. gr. 7.3.—Fuses at 442°. Lustre white, brilliant, and little affected by air or moisture. Very malleable and ductile; produces a crackling noise when bent. Burns at a white heat, forming peroxide of tin.

1053. When required in a minute state of division it is melted and poured into a wooden box, previously rubbed over with chalk. By brisk agitation it is reduced to a state of powder. On putting it through a sieve, and floating away any adhering particles of chalk with water, it is obtained in the form of a very fine powder.

1054. Prepared from the native oxide, its oxygen being removed by charcoal. The purer kinds of tin are usually called grain tin, and the other block tin.

1055. When tin-plate is heated and washed with a liquid made up of 8 parts of water, 1 of nitric, and 1 of muriatic acid, by measure, the surface assumes a crystalline appearance, the external layer of particles being dissolved by the acid fluid. In this state it forms the moiree metallique, or crystallized tin-plate.

1056. Oxide or Protoxide of Tin.—Symb. ·Sn = tin 58 + oxygen 8.—

Eq. 66.—Insoluble in water, dissolved by potassa, burns like tinder when heated in the air. Has a great attraction for oxygen and chlorine; its salts precipitate many metals from their solutions.

1057. BINOXIDE or PEROXIDE OF TIN.—Symb. :Sn = tin 58 + oxygen 16.—Eq. 74.—Insoluble in water, soluble in a solution of potassa. Insoluble in nitric acid after it has been heated. Colour yellowish-white. Its salts have no deoxidating properties like those of the oxide.

1058. PREPARED by heating tin or oxide of tin freely in the open air; or by adding a solution of potassa to permuriate of tin.

1059. Sulphuret of Tin.—Symb. SSn.—Formed by heating tin and sulphur, or by the action of sulphureted hydrogen on oxide of tin in solution. See 711.

1060. Bisulphuret of tin is prepared by heating equal weights of sulphur and peroxide of tin; 1 part of muriate of ammonia added to 4 of the mixture, promotes its production. It forms a crystalline scaly mass, and from its colour and lustre is often called *mosaic gold.*

1061. MURIATE OF TIN in solution is much used as a deoxidating agent. Prepared by heating tin in common muriatic acid so long as hydrogen gas is evolved. The hydrogen disengaged has an offensive odour, from impurities which it derives from the tin.

1062. BIPERMURIATE OF TIN is procured by the action of nitric and muriatic acids on tin; a violent reaction takes place unless the tin be added in small portions at a time. Procured also by exposing a solution of the muriate to the action of the air for a length of time.

1063. BICHLORIDE OF TIN or FUMING LIQUOR OF LIBAVIUS, is obtained by distillation from 1 part of mercury, 6 of tin, and 33 of bichloride of mercury.

1064. *Tests of Tin in solution.*—The oxide precipitates solutions of gold, platinum and mercury. Hydrosulphuric acid gives a brown precipitate. Potassa precipitates the oxide, and an excess dissolves it. The peroxide gives a yellowish-coloured precipitate with hydrosulphuric acid.

CHAP. VII. BISMUTH.

1065. Symb. Bi.—Eq. 72,—Sp. gr. 9.85.—Melts at 496°, and volatilized by a strong heat. Brittle, highly crystalline; colour white, with a shade of red.

1066. OXIDE OF BISMUTH.—Symb. ·Bi = bismuth 72 + oxygen 8.—Eq. 80.—Colour yellow; fusible. Prepared by heating bismuth in contact with air, or by the action of potassa with the nitrate of bismuth, the nitric acid being removed by the potassa and the oxide precipitated.

1067. TRINITRATE OF BISMUTH, called also SUBNITRATE OR MAGISTERY OF BISMUTH.—Symb. ::N+3·Bi.—White, insipid, insoluble in water.

1068. PREPARED by pouring a solution of nitrate of bismuth into a large quantity of water, being immediately precipitated; most of the acid, with a little oxide, remains in solution.

1069. NITRATE OF BISMUTH prepared by the action of diluted nitric acid with bismuth. The metal is oxidated by part of the acid, nitric oxide gas being evolved, and the oxide combines with the undecomposed acid.

1070. *Tests of Bismuth.*—Distinguished by the nitrate being so easily decomposed with water, and the dark coloured sulphuret which the oxide gives with hydrosulphuric acid (sulphureted hydrogen, see 711); the oxide precipitated by potassa is reduced with great facility on charcoal by heat, when the metal is easily recognised by its fusibility and brittleness.

CHAP. VIII. MANGANESE.

1071. Symb. Mn.—Eq. 28.—Similar in appearance to iron. Brittle, fused with difficulty, attracts oxygen from air and water. PREPARED by decomposing the oxide with charcoal in a crucible lined with charcoal.

1072. *Oxides of Manganese.*—OXIDE or PROTOXIDE.—Symb. ·Mn = manganese 28 + oxygen 8.—Eq. 36.—Colour green, attracts oxygen rapidly when heated, forms colourless salts with the acids. Prepared by transmitting hydrogen gas over any of the other oxides heated in a tube.

1073. BINOXIDE or PEROXIDE.—Symb. :Mn = manganese 28 + oxygen 8.—Eq. 36.—Found native in abundance. Used principally in the preparation of oxygen, chlorine, and flint-glass.

1074. MANGANESIOUS ACID.—Symb. ⁝Mn.

1075. MANGANESIC ACID.—Symb. ⁞Mn.—When peroxide of manganese and nitre are heated, a compound is obtained called the mineral CHAMELEON. It consists of manganesious acid and potassa, gives a green coloured solution in water, which gradually becomes purple and red on exposure to the air, after passing through many intermediate tints; the red coloured compound is formed by the manganesious compound attracting oxygen from the air or water.

1076. DEUTOXIDE OF MANGANESE.—Symb. ⁝Mn^2.—Eq. 80.—Formed when the peroxide is exposed to heat, as in the preparation of oxygen gas.

1077. RED OXIDE OF MANGANESE.—Symb. ⁞Mn^3.

1078. VARVICITE.—Symb. ⁞Mn^7.

1079. Sulphate and muriate of manganese are formed during the preparation of oxygen and chlorine gases.

1080. Very pure muriate of manganese is procured when muriate of ammonia is heated with peroxide of manganese, dissolving it afterwards in water.

Tests of Manganese.

1081. FERROPRUSSIATE OF POTASSA gives a white precipitate. The oxide precipitated by alkalis gives an amethystine tint to borax heated at the blowpipe

1082. Hydrosulphuret of Ammonia gives a pale rose-coloured precipitate.

CHAP. IX. CHROMIUM.

1083. Symb. Cr.—Eq. 28?—Sp. gr. 6.—Brittle, fused with difficulty. Colour white, with a slight yellow tint. PREPARED by heating intensely oxide of chromium mixed with charcoal.

1084. OXIDE OF CHROMIUM (PROTOXIDE).—Symb. ·Cr = chromium 28 + oxygen 8.—Eq. 36.—Colour green; combines with acids, forming salts, which are generally of a greenish tint.

1085. PREPARED by heating CHROMATE OF MERCURY to redness in a crucible The oxide of mercury is decomposed and expelled, along with part of the oxygen of the chromic acid, the oxide of chrome being left. When salts containing chromic acid are heated with sulphuric acid, and any deoxidating substance, as alcohol or sugar, a green tint is generally produced, oxide of chrome being formed.

1086. CARBONATE OF CHROMIUM has a fine greenish-blue colour. It is prepared by boiling bichromate of potassa in solution with sugar, separating the brown matter that is precipitated, and boiling it with carbonate of potassa (Mr T. Thomson).

1087. CHROMIC ACID.—Symb. ⁝Cr = chromium 28 + oxygen 24.—Eq. 52.—

Colour deep red. Deliquescent, soluble in water, crystallizable. Decomposes muriatic acid, and liberates chlorine; the resulting fluid dissolves gold in the same manner as the nitromuriatic acid.

1088. PREPARED by heating fluor-spar, chromate of lead, and aqueous sulphuric acid, passing the gaseous compound so procured into water, and heating it in a platinum crucible. The acid and fluor-spar give hydrofluoric (698) and chromic acids with the water, the hydrofluoric acid being expelled and the chromic left.

Compounds of Chromic Acid.

1089. CHROMATE OF POTASSA.—Colour yellow, very soluble, crystallizable. Prepared by heating chrome iron-ore with nitre, dissolving the resulting soluble matter, and crystallizing. The nitre converts the chrome into chromic acid by communicating oxygen, and this uniting with the potassa of the nitre produces the chromate. The iron is left as a peroxide.

1090. BICHROMATE OF POTASSA.—Colour ruby red; soluble in 10 parts of cold water, crystallizable. Formed when the solution of the chromate is mixed with sulphuric or some other acid, which combines with part of the potassa, crystallizing by spontaneous evaporation.

1091. CHROMATE OF LEAD.—Colour brilliant yellow. Insoluble. PRECIPITATED when chromate or bichromate of potassa mixes in solution with a salt of lead.

1092. DICHROMATE OF LEAD, commonly called SUBCHROMATE OF LEAD.—Prepared by boiling the chromate of lead in a solution of potassa or lime-water, which removes one-half of the chromic acid, and leaves the dichromate of a fine orange colour.

1093. CHROMATE OF MERCURY.—Orange-red colour. PRECIPITATED when a solution containing chromate of potassa is mixed with a solution of nitrate of mercury (nitrate of the protoxide).

1094. CHROMATE OF SILVER has a purplish tint; it may be procured by precipitation on mixing solutions of chromate of potassa and nitrate of silver.

1095. CHLOROCHROMIC ACID.—When aqueous sulphuric acid is mixed with common salt, muriatic acid gas is evolved (647) and sulphate of soda left; if bichromate of potassa and an additional quantity of sulphuric acid be added to the mixture for producing muriatic acid, sulphate of potassa is formed, and chromic acid disengaged. The muriatic and chromic acids produce the chlorochromic acid; and as they are believed by some chemists to decompose each other completely, all the oxygen of the chromic acid being removed by hydrogen from the muriatic acid, the compound that results may be more correctly termed TERCHLORIDE OF CHROMIUM. It is very volatile; fluid; has a brilliant ruby colour; decomposes water, producing chromic and muriatic acids. It acts violently with numerous substances containing much hydrogen, as alcohol, ether, and ammonia; the chlorine combines with the hydrogen, forming muriatic acid, and the chromium, taking oxygen, forms oxide of chrome. Heat and light are evolved during the reaction.

1096. FLUOCHROMIC ACID is a similar compound, prepared by the action of aqueous sulphuric acid with chromate of lead and fluor-spar, the same changes taking place between the hydrofluoric and chromic acids which are evolved, as between the muriatic and chromic acids, according to the explanation given in the preceding paragraph. TERFLUORIDE OF CHROMIUM may therefore be considered a more appropriate name.

CHAP. X. COBALT.

1097. Symb. Co.—Eq. 26.—Sp. gr. 8.5.—Brittle; colour gray, with a reddish tint. Melts at a temperature a little lower than is necessary to fuse iron. Attracted by the magnet, and may be rendered permanently magnetic like iron.

1098. Prepared by heating the oxalate of cobalt, carbonic acid being disengaged as the oxide and the oxalic acid are decomposed, metallic cobalt being left. The oxalate is procured by the action of oxalic acid on the carbonate, which is precipitated when the carbonate of potassa is added to the muriate of cobalt.

1099. Oxide of Cobalt.—Symb. ·Co = cobalt 26 + oxygen 8.—The oxide is gray when dry, but if it be precipitated from solutions of its salts by potassa, it has a bluish colour, a portion of water being precipitated along with it. The salts of cobalt have usually a reddish colour. Other oxides have been described, but they are of less importance. Zaffre is the impure oxide that remains when the arseniuret of cobalt has been heated to expel arsenic, the cobalt being, at the same time, oxidated. When this oxide is fused with siliceous matter, it forms the blue coloured vitreous compound named smalt, much used in colouring glass blue.

1100. Muriate of Cobalt is soluble, of a reddish colour in solution, but blue when dry. It is used as a sympathetic ink; traces, invisible at first, becoming blue by heat. If mixed with common salt, traces drawn with the solution become of a fine green when heated. The change of colour continues only so long as the heat is continued, and the material kept quite dry. Nitromuriatic acid is generally used as a solvent of cobalt.

CHAP. XI. ARSENIC.

1101. Symb. As.—Eq. 38.—Sp. gr. 5.7. Steel-gray colour, brilliant lustre, but soon tarnishes on exposure to the air. Volatilizes at 360°, producing a garlic odour; inflames when subjected to a high temperature, and freely exposed to the air, producing arsenious acid.

1102. Prepared by heating arsenious acid with charcoal; the carbon takes the oxygen of the acid, and the metal is converted into vapour, which must be protected from the air as it is condensed. On the small scale, a tube is employed; when larger quantities are used, the mixture is put into a crucible, and another crucible inverted over it to condense the metal.

1103. Oxide or Protoxide of Arsenic.—The black powder that is formed when some varieties of arsenic are exposed to the air. Regarded by many as a mixture of arsenic and the white oxide.

1104. Arsenious Acid, White Oxide, Binoxide, or Deutoxide of Arsenic.—Symb. :As = arsenic 38 + oxygen 16.—Eq. 54.—Sp. gr. 3.7.—Obtained in the form of solid, semitransparent masses, which become opaque by keeping, and exposure to the air. Volatilizes at 380°; produces no garlic odour; condenses in small octohedral crystals, which have a brilliant adamantine lustre. At a red heat it is resolved into metallic arsenic and arsenic acid. Soluble in 16 of water at 212°, 5–8ths being again deposited in crystals as it cools. Cold water mixed with arsenious acid takes up only 1–100th part. The presence of organic substances diminishes its solubility. Has little or no taste, but if kept long on the tongue it produces an acrid sensation, inducing inflammation. (Dr Christison.) It acts feebly on the vegetable blues; but litmus is immediately reddened by it.

1105. Prepared during the roasting of ores of cobalt and other metals, the metallic arsenic being volatilized by the heat, oxidating in contact with air, and condensing into solid masses in flues connected with the furnaces. Arsenious acid is the poison commonly known by the name of Arsenic. It is recognised in the following manner:—

1106. I. WHEN SOLID.—By the crust of metallic arsenic which it gives when heated in a tube with charcoal, and the various properties described in 1101, which the metallic crust presents. The black flux is often used instead of charcoal.

1107. II. WHEN IN SOLUTION.—By passing HYDROSULPHURIC ACID (sulphureted hydrogen) through the solution, drying the yellow compound of sulphur and arsenic which is thus precipitated, and heating it with dry carbonate of soda, when a crust of metallic arsenic is procured, possessing the properties referred to in the preceding paragraph.

1108. In this process the oxygen of the arsenious acid unites with the hydrogen of the hydrosulphuric acid, and a yellow sulphuret of arsenic is precipitated; the carbonate of soda is decomposed by the sulphuret, carbonic acid and oxygen escaping, while the sulphur combines with the sodium. The metallic arsenic is then sublimed; the whole of the arsenic is not procured, however, as a portion of the metal remains combined with part of the sodium.

1109. AN EXCESS of HYDROSULPHURIC ACID must be avoided, as it would prevent a portion of the yellow sulphuret from being precipitated. By boiling the liquid any excess is expelled.

1110. Any FREE ALKALI in solution must be neutralized; it would prevent the precipitation of the yellow sulphuret. Muriatic or acetic acid is generally used, and added till slightly in excess.

1111. Dr Christison has recommended the process with hydrosulphuric acid as the best, and pointed out numerous important circumstances to be attended to in conducting it. He states that the metallic arsenic sublimed in a small tube may be recognised by its appearance, even when it does not exceed the 280th part of a grain.

1112. When the quantity of metal is very minute, and not very distinct, it may be oxidated by heating it in the tube, when arsenious acid may often be seen in brilliant crystals; and after touching them with a drop of water, the following liquid tests may be applied.

1113. A drop of AMMONIACO-NITRATE OF SILVER added to a very small portion, produces a yellow precipitate of arsenite of silver. A drop of AMMONIACO-SULPHATE OF COPPER, added in the same way, produces a grass-green compound, arsenite of copper.

1114. These AMMONIACAL TESTS are prepared by adding ammonia to solutions of nitrate of silver and sulphate of copper, until the oxide, precipitated at first, shall have been almost entirely redissolved. An excess of ammonia must be avoided, the arsenites of copper and silver being both soluble in an excess of this alkali. Instead of using the ammoniacal salts, sulphate of copper and nitrate of silver may be added alone, and a small portion of potassa or soda introduced afterwards to detach the acid from these salts, and allow the arsenious acid to act upon the oxides of silver and copper.

1115. Arsenious acid has too feeble an attraction for these oxides to separate them from any of the stronger acids.

1116. In examining liquids, suspected to contain arsenious acid, with ammoniaco-nitrate of silver, it must be recollected that PHOSPHATE OF SODA in solution precipitates phosphate of silver, similar in colour to arsenite of silver; but the precipitate is smooth and uniform, instead of curdy like the arsenite of silver.

1117. Again, if much MURIATIC ACID, or common salt, be present, along with arsenious acid, there will be so much chloride of silver precipitated, that it will be impossible to observe if any arsenite of silver be at the same time formed, without the aid of another process.

1118. In using the salt of copper, the blue precipitate that appears on the addition of an alkali, may be rendered green from other causes than the presence of arsenious acid, as a yellow colour in the fluid under examination, through which the blue hydrated oxide will appear green like arsenite of copper. Hence, as none of the silver or copper tests can individually be relied upon, the process with sulphureted hydrogen is preferred.

1119. Mr E. Davy recommends platinum to be used in precipitating arsenic, touching it with zinc. The arsenic adheres to the platinum, but may be separated by heat.

1120. Many other substances have been recommended as tests of arsenious acid, but they are not so delicate in their operation, and they are so liable to fallacy from the numerous substances so frequently found in mixed solutions, that they can seldom be depended upon in endeavouring to detect arsenious acid in such liquids.

1121. LIME-WATER gives a white precipitate with arsenious acid.

1122. BICHROMATE OF POTASSA in solution becomes slowly green, from the deoxidation of part of the chromic acid.

1123. NITRATE OF POTASSA may be used occasionally with advantage where the suspected matter is evidently mixed with a considerable quantity of animal or vegetable matter. Equal parts of nitre and the suspected matter are heated together; the nitre destroys any organic matter, and, at the same time, converts any arsenious acid that may be present into arsenic acid, which unites with the potassa. The arseniate of potassa thus produced is dissolved readily by water; and nitrate of silver added to the solution gives a BRICK RED precipitate of ARSENIATE OF SILVER.

1124. Astringent solutions, vinegar, sulphureted hydrogen water, alkaline hydrosulphurets, and other substances, have been proposed as ANTIDOTES TO ARSENIC, but no dependence is placed upon them. Of late, however, the peroxide of iron has been recommended for this purpose, producing arseniate of iron, which is inert and insoluble. The hydrated peroxide is prepared for this purpose from a persalt of iron by ammonia, which immediately precipitates it, uniting with the acid, and remaining in solution.

1125. ARSENITE OF POTASSA in solution is formed when arsenious acid is boiled in water with an equal weight of carbonate of potassa; 4 grains of arsenious acid are contained in each ounce of fluid; it is usually called FOWLER'S SOLUTION, or LIQUOR ARSENICALIS.

1126. ARSENIC ACID, or PEROXIDE OF ARSENIC, may be obtained by heating arsenious acid with nitric acid, and evaporating till a syrupy fluid is procured, which must be afterwards heated to redness to expell any excess of acid. If the arsenious acid be first dissolved in muriatic acid, and then treated as above, arsenic acid is procured more readily.

1127. ARSENIC combines with CHLORINE, when thrown in a minute state of division into this gas. Heat and light are evolved during the combination.

1128. HYDROGEN and ARSENIC combine in different proportions. The compound that has attracted most attention is the ARSENIURETED HYDROGEN GAS, evolved when an alloy of tin and arsenic is dissolved in muriatic acid. It is colourless, inflammable, has an offensive odour, and is exceedingly poisonous.

1129. SULPHUR and ARSENIC combine in three or four proportions.

1130. SULPHURET OF ARSENIC (PROTOSULPHURET), or REALGAR, is formed by heating arsenious acid with 2-5ths of its weight of sulphur, the oxygen of the arsenious acid being removed by part of the sulphur. It has a reddish colour.

1131. ORPIMENT, YELLOW SULPHURET OF ARSENIC, or KING'S YELLOW, is formed when equal weights of sulphur and arsenious acid are heated, or by the action of hydrosulphuric acid on arsenious acid in solution. The other compounds are unimportant.

CHAP. XII. NICKEL, VANADIUM, &c.

1132. The remaining metals belonging to this class are very rarely met with, compared with these which have been already described.

1133. NICKEL.—Symb. Ni.—Eq. 26.—Sp. gr. 8.2.—White, malleable, ductile, attracted by the magnet, and may be rendered magnetic like iron. Fused at an intense heat, and though placed among the metals whose oxides are reduced by heat alone, by some chemists, it appears that the propriety of this arrangement is doubtful, and many have placed it amongst the metals of the preceding class.

1134. The nitric and nitromuriatic acids are the best solvents of nickel. Solutions of its salts have a fine green colour, and give an apple-green precipitate of the oxide when decomposed by potassa. Ammonia in excess redissolves the oxide, and gives a blue coloured solution, similar, but not so deep, as that produced by solutions of copper with ammonia.

1135. VANADIUM is procured from some ores of iron and lead.

1136. CADMIUM is usually obtained from ores of zinc.

1137. TUNGSTEN or WOLFRAM is a very heavy metal, which is prepared from tungstic acid.

1138. MOLYBDENUM is obtained from the native sulphuret, which resembles plumbago in appearance.

1139. CERIUM is found in the minerals called Cerite and Allanite.

1140. COLUMBIUM or TANTALUM was discovered in the Tantalite, Ittrotantalite, and other minerals.

1141. TELLURIUM has been extracted principally from some ores of gold and silver.

1142. TITANIUM has been procured from Titanite, and also from the slag separated during the preparation of iron from some of its ores.

1143. URANIUM is prepared from pitch blende, which contains oxides of titanium and of iron.

2. COMMON METALS WHOSE OXIDES ARE REDUCED BY HEAT ALONE.

CHAP. I. MERCURY OR QUICKSILVER.

1144. Symb. Hg.—Eq. 200.—Sp. gr. 13.5.—Freezes at —39.5°, contracting much during congelation; malleable when solid. Boils at 662° (650° according to other estimates). Liquid, brilliant white metallic lustre, forms round globules, which divide readily and coalesce with facility when brought in contact. Tarnishes quickly when exposed to the air if impure, small portions of other metals dissolved in it being oxidated, and forming a crust at its surface.

1145. PREPARED from CINNABAR, the native bisulphuret of mercury, which, on the small scale, is heated with lime, iron, or carbonate of potassa or soda; these remove the sulphur, and the mercury is separated by distillation.

4

Fig. 44.

Materials.		Products.
232 Bisulphuret of Mercury.........	Mercury 200	200 Mercury.
	Sulphur... 16	
	Sulphur... 16	
Iron...................................	28	44 Sulphuret of Iron.
Iron...................................	28	44 Sulphuret of Iron.

Symb. 2S+Hg&2Fe = Hg&2SFe.

1146. Purified from any metals which may have been dissolved in it, by slow distillation, pouring over it afterwards sulphate or nitrate of mercury in solution, previously mixed with an excess of acid. The metals with which it is commonly adulterated are lead, tin, bismuth, and zinc, and a minute portion of any of these which may have been carried over accidentally during the distillation, is oxidated and separated from the pure mercury by the salt employed.

1147. Oxide (or Protoxide) of Mercury, commonly called Gray, Ash, or Black Coloured Oxide of Mercury.—Symb. ·Hg = mercury 200 + oxygen 8.—Eq. 208.—Insoluble in water and muriatic acid, dissolved by acetic acid. Prepared by adding potassa, soda, or lime-water, to salts containing oxide of mercury; formed also by mixing lime-water with the chloride of mercury. The hydrogen of part of the water unites with the chlorine and produces muriatic acid, which remains in solution with the lime, while the oxygen and mercury combine to form the oxide. By long keeping and exposure to the light, it is apt to pass into metallic mercury and peroxide of mercury.

Fig. 45.

Materials.		Products.
28 Lime........................	28	65 Muriate of Lime.
9 Water	Hydrogen 1	
	Oxygen... 8	
236 Chloride of Mercury ...	Chlorine 36	
	Mercury 200	208 Protoxide of Mercury.

Symb. ClHg&·H&·Ca = HCl+·Ca&·Hg.

1148. Mercury treated with manna or lard, till the globules disappear, has been supposed to be oxidated, but many are now of opinion that the mercury is still in the metallic form, and merely in a minute state of division.

1149. Binoxide or Peroxide of Mercury.—:Hg = mercury 200 + oxygen 16.—Eq. 216.—Red, shining, crystalline, very sparingly soluble in water, and commonly said to be insoluble; turns feebly to a green some of the vegetable blues. When precipitated from its aqueous solutions, it occasionally presents a yellow colour from the presence of water in combination.

1150. Prepared on decomposing bipernitrate of mercury by heat, all the nitric acid, with the exception of a very minute quantity, being decomposed and expelled. It is of a dark colour when hot, but appears in the form of brilliant red scales as it cools; it is often called Red Precipitate when procured in this manner. Prepared also by keeping mercury for some weeks at a temperature about 608°, when it is slowly oxidated. A higher temperature decomposes the oxide; formed by this process it is called Precipitate per se (by itself).

1151. Nitric Acid and Mercury.—Several compounds of nitric acid and mercury have been described. Mercury is oxidated by nitric acid rapidly, and then combines with a portion of undecomposed acid. Where the acid is in excess, not diluted, and the action assisted by heat, peroxide of mercury is always formed, united with a large quantity of acid; the compound may be termed Bipernitrate of Mercury, supposing it to contain 2 eqs. of acid and 1 eq. of peroxide of mercury. But if the mercury be in excess, the acid diluted, and the action allowed to proceed at natural temperatures, oxide of mercury is alone obtained in combination with the acid; and if 1 eq. of acid be combined with the oxide, then the compound is termed simply NITRATE OF MERCURY.

1152. If a solution of bipernitrate of mercury be thrown into a large quantity of water, a precipitate of a yellowish colour is formed, regarded by many chemists as a DIPERNITRATE OF MERCURY = 2 eqs. of peroxide + 1 eq. of acid. There is at present a considerable difference of opinion as to the composition of various salts of nitric acid and mercury.

1153. BIPERSULPHATE OF MERCURY.—Symb. 2⁝S+:Hg = 2 eqs. sulphuric acid 80 + 1 peroxide 216.—Eq. 296.—White, crystalline. Employed principally in the preparation of bichloride, sulphate, and persulphate of mercury.

1154. PREPARED by boiling 2 parts of mercury with 3 of aqueous sulphuric acid to dryness. 2 eqs. of acid each lose 1 eq. of oxygen, which convert the mercury into peroxide, sulphurous acid being evolved; other 2 eqs. of acid unite with the peroxide to form the salt required. The excess of acid is expelled.

Fig. 46.

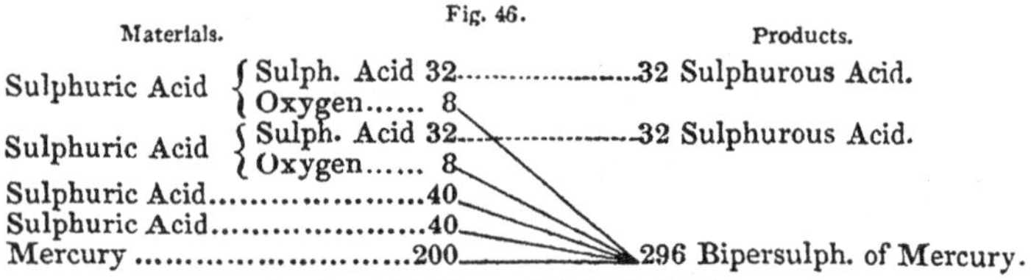

Symb. 4⁝S&Hg = 2⁝S+:Hg&2:S.

1155. SULPHATE OF MERCURY.—Symb. ⁝S·Hg = sulphuric acid 40 + oxide of mercury 208.—Eq. 248.—Formed, according to the statements of many chemists, when mercury is heated very gently with sulphuric acid, or bipersulphate of mercury is mixed intimately with metallic mercury.

Fig. 47.

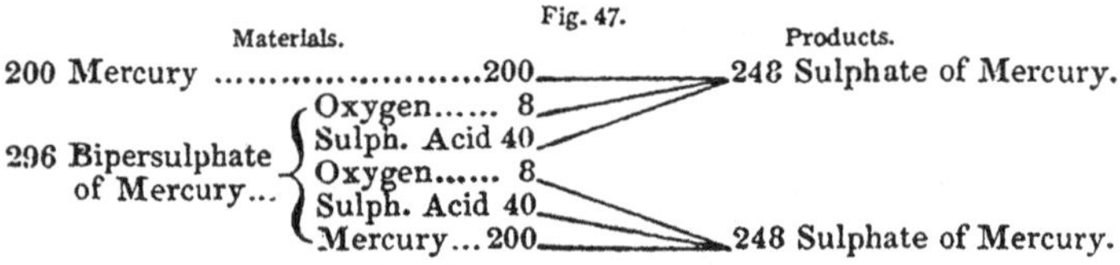

Symb. 2⁝S+:Hg&Hg = 2⁝S·Hg.

1156. PERSULPHATE OF MERCURY.—Symb. ⁝S:Hg = sulphuric acid 40 + peroxide of mercury 216. Commonly called Turpeth Mineral.—Precipitated of a yellow colour when bipersulphate of mercury is thrown into boiling water. Mr Phillips states, that it consists of 3 eqs. of acid + 4 eqs of peroxide, according to his experiments.

1157. BISULPHURET OF MERCURY.—Symb. S^2Hg = sulphur 32 + mercury 200.—Eq. 232.—It constitutes VERMILION, when prepared with much care, and reduced to a minute state of division to present the proper vermilion tint. Called CINNABAR as it is met with in nature, and FACTITIOUS CINNABAR when prepared artificially. Crystalline, and of a dull vermilion tint, when first prepared and sublimed. Formed generally by adding mercury to melted sulphur, and subliming the dark coloured compound which is produced. Precipitated of a black colour from solutions of salts of mercury by sulphureted hydrogen.

Fig. 48.

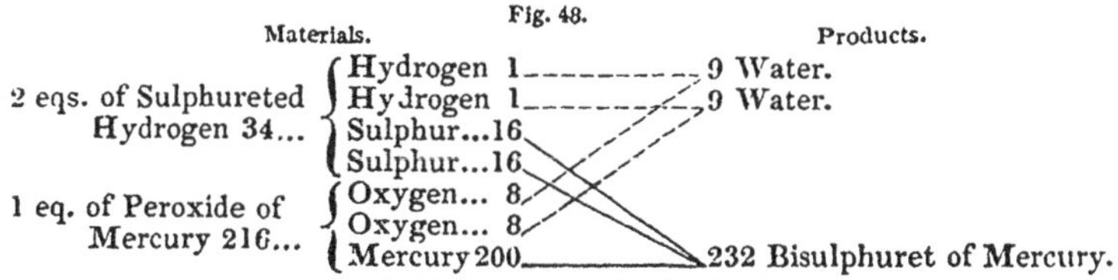

Symb. 2HS&:Hg = 2·H&2S+Hg.

1158. Ethiops Mineral is prepared by triturating equal weights of sulphur and mercury, till the metallic globules disappear. It is composed of bisulphuret of mercury and sulphur according to Brande.

1159. Carbonic Acid has a very feeble attraction for the oxides of mercury; some compounds of carbonic acid and the peroxide have been described, but the oxide and carbonic acid cannot unite according to the opinion now entertained.

1160. Acetate of Mercury is prepared by mixing the nitrate of mercury with a hot solution of the acetate of potassa; it is deposited in small scaly crystals as the liquid cools.

Fig. 49.

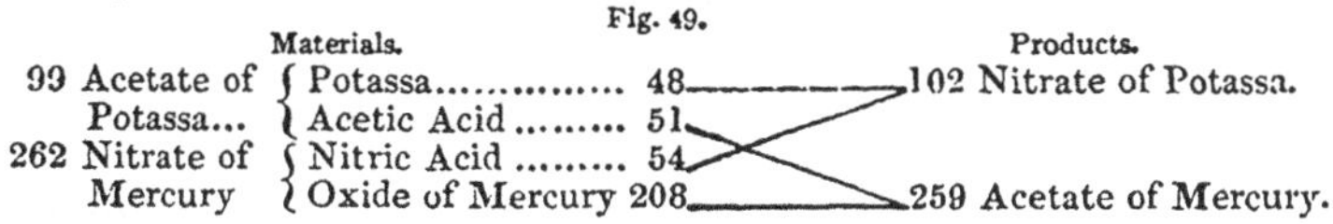

Symb. *A*·K&::N·Hg = *A*·Hg&::N·K.

1161. Chloride of Mercury, or Calomel; formerly called Submuriate (or Mild Muriate) of Mercury.—Symb. ClHg = chlorine 36 + mercury 200.—Eq. 236.—Solid, crystalline, insoluble, tasteless. Converted into vapour at a temperature below redness, but does not melt when heated. Decomposed by solutions of alkalis and alkaline earths, producing the oxide of mercury and muriatic acid.

1162. Preparation. By precipitation from nitrate of mercury and chloride of sodium, mixed together in solution, the chloride of mercury being deposited and nitrate of soda remaining in solution.

Fig. 50.

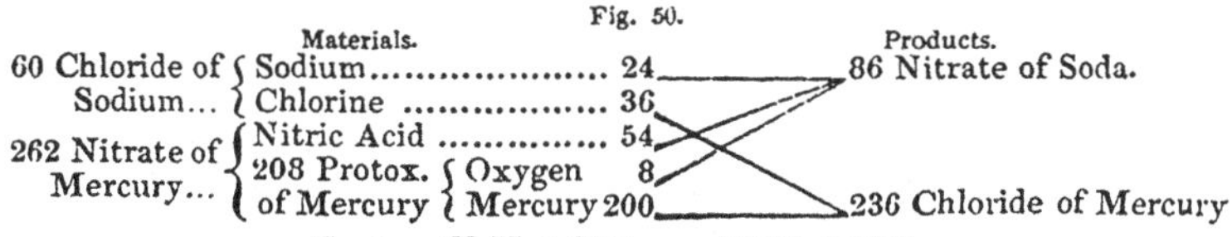

Symb. ::N·Hg&ClNa = ::N·Na&ClHg.

1163. In preparing the precipitated chloride, if any bipernitrate of mercury be present, it produces bichloride of mercury instead of chloride, which remains in solution with the nitrate of soda, as in the following diagram :—

Fig. 51.

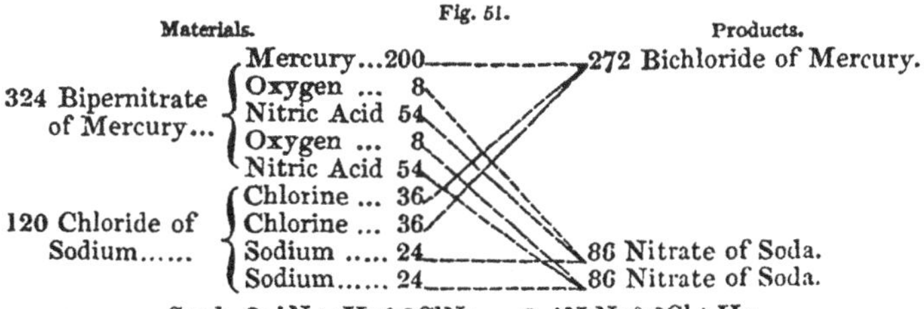

Symb. 2::N+:Hg&2ClNa = 2::N·Na&2Cl+Hg.

1164. By SUBLIMATION from a mixture of 1 eq. of the sulphate of mercury and 1 eq. of chloride of sodium, the products being 1 eq. of chloride of mercury and 1 eq. of sulphate of soda.

Fig. 52.

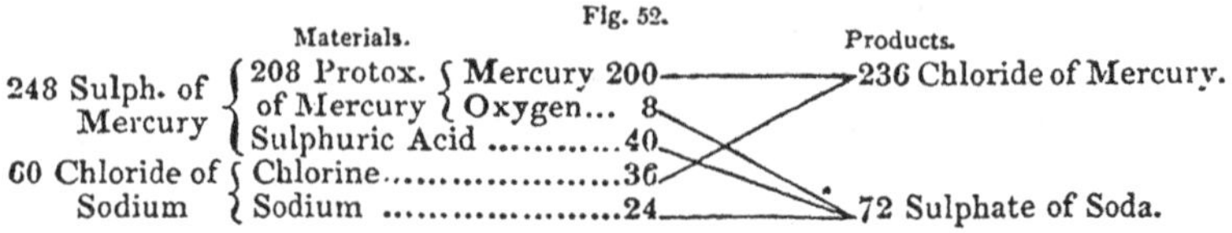

Symb. ⁝S·Hg&ClNa = ⁝S·Na&ClHg.

1165. By SUBLIMATION from a mixture of 1 eq. of bichloride of mercury and 1 eq. of mercury, when 2 eqs. of calomel are formed.

Fig. 53.

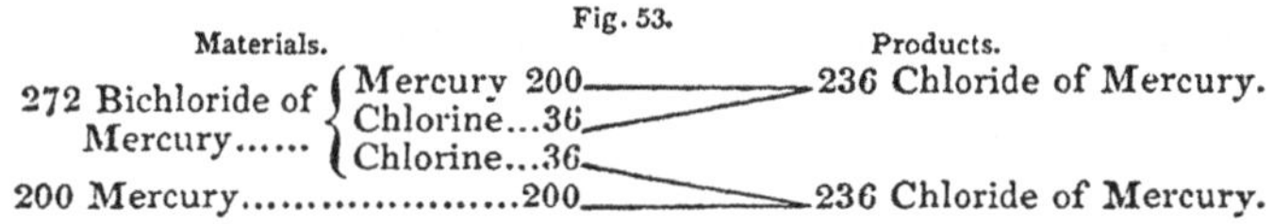

Symb. 2Cl+Hg&Hg = 2ClHg.

1166. If mercury be brought in contact with chlorine at natural temperatures, the chlorine is absorbed, and a similar compound formed.

1167. Calomel is generally mixed with a little bichloride of mercury when first prepared, which is removed by water.

1168. BICHLORIDE OF MERCURY, or CORROSIVE SUBLIMATE.—Symb. Cl^2Hg. —Eq. 272.——Solid, semitransparent, crystalline; converted easily into vapour by heat, condensing into a cake, or prismatic crystals. Soluble in 20 of cold, and in about 3 of boiling water; soluble also in alcohol and ether. 2 eqs. of water are decomposed by 1 eq. of bichloride, 2 of muriatic acid and 1 of peroxide being formed, which remain in union. But when decomposed by solutions of fixed alkalis, and alkaline earths, peroxide of mercury is precipitated, while compounds of muriatic acid and the precipitating agent remain in solution.

Fig. 54.

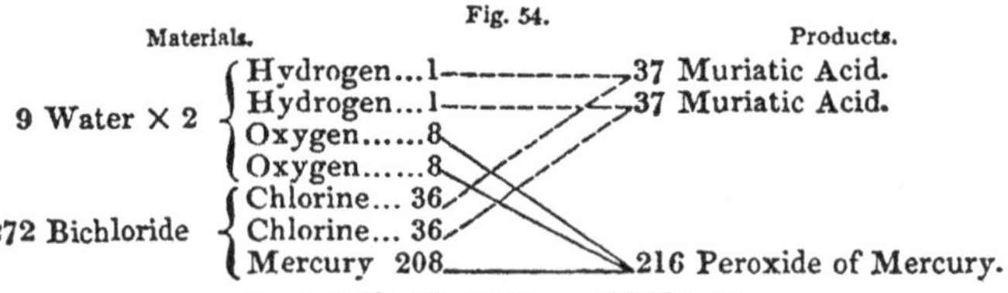

Symb. 2Cl+Hg&2·H = 2HCl+:Hg.

1169. The following diagram explains the manner in which the bipermuriate may be again converted into bichloride and water :—

Fig. 55.

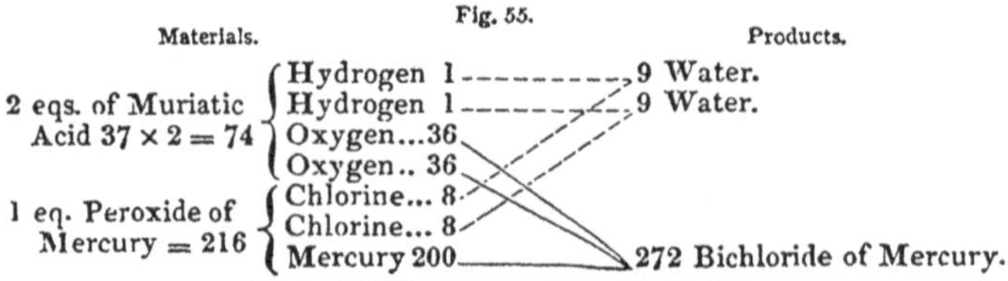

Symb. 2HCl+:Hg = 2Cl+Hg&2·H.

1170. Albumen precipitates calomel, or binoxide of mercury according to recent experiments, combining with it, and forming an inert compound; hence albumen is used as an antidote to bichloride of mercury.

1171. **Prepared** by heating 1 eq. of bipersulphate of mercury with 2 of chloride of sodium (common salt); 2 eqs. of sulphate of soda are formed, and 1 of bichloride of mercury is sublimed.

Fig. 56

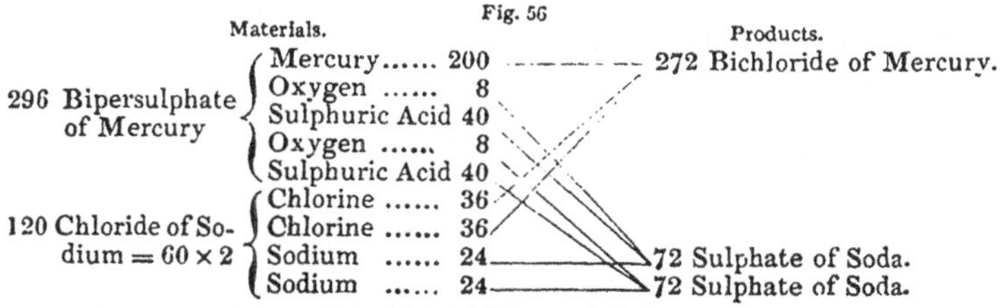

Symb. 2⁝S+:Hg&2ClNa = 2⁝S·Na&2Cl+Hg.

1172. **Muriate of Ammonia and Mercury** has been regarded as a compound of muriatic acid, ammonia, and peroxide of mercury; but Dr Kane conjectures that it is a compound of bichloride of mercury and binamide of mercury, the binamide being formed by the decomposition of ammonia, and represented by $2NH^2+Hg$. It is formed when ammonia is added to a solution of the bichloride of mercury, or when potassa or its carbonate is added to the bichloride previously dissolved in a solution of muriate of ammonia. It is often called SAL ALEMBROTH, or the WHITE precipitate of mercury.

1173. **Iodide of Mercury.**—Symb. IHg = iodine 124 + mercury 200.—Eq. 324.—**Precipitated** as a yellow powder (with a tint of green) when nitrate of mercury is mixed with a solution of iodide of potassium, nitrate of potassa remaining in solution.

1174. **Biniodide or Periodide of Mercury.**—Symb. I^2Hg = iodine 248 + mercury 200.—Eq. 448.—**Precipitated** when bipernitrate or bichloride of mercury in solution is added to a solution of iodide of potassium. Formed, also, when iodine and mercury are heated together. It has a fine yellow or crimson colour; the yellow periodide often turns of a crimson colour merely on touching it, and the progress of the crimson tint through the whole mass may be distinctly traced.

Tests of Mercury.

1175. **Precipitated** from solutions in the metallic form by COPPER, TIN, and other metals; by some deoxidating salts, as MURIATE OF TIN; and by GOLD and PLATINUM, when a drop of the solution is placed upon them, and they are touched at the part where it rests with iron, tin, or some other very oxidable metal. The precipitated matter gives metallic globules by heat.

1176. **Hydrosulphuric Acid** in excess gives a dark bisulphuret, which may be converted into vermilion by heat.

1177. Lime-water and potassa precipitate oxide or peroxide of mercury. Potassa in excess is often used to destroy adhering organic matter, and the metallic mercury procured by heat. When bichloride of mercury is suspected in solutions containing merely organic matter, Dr Christison recommends the bichloride to be removed by sulphuric ether.

CHAP. II. SILVER.

1178. Symb. Ag.—Eq. 110.—Has a brilliant white colour, and great lustre when polished. Melts at a bright red heat, absorbing oxygen when pure, and evolving it again as it cools. Extremely ductile and malleable.

1179. Silver is procured from its ores by different processes, according to the condition in which it is found. When combined with lead which has been prepared from galena, the lead is oxidated by the action of heat and air, and the melted oxide removed, while the silver is left in a pure form. If the silver be combined with sulphur, it is separated by the process of AMALGAMATION. For this purpose the ore is mixed with common salt, and exposed to air and heat. Chloride of silver is thus procured, along with sulphate of soda. The mixture is reduced to powder, and agitated with water, iron, and mercury; the iron removes the chlorine, and the mercury unites with the silver. The amalgam procured is decomposed by heat, after the mercury mechanically mixed with it has been separated; the mercury being volatilized and the silver left.

1180. When silver is purchased for experimental purposes, it is usually found alloyed with copper. The copper may be separated by any of the following methods:—

1181. I. Dissolve the alloy in nitric acid, and precipitate the silver by common salt, as chloride of silver; the copper is left in solution. Wash the chloride with water, and heat it with carbonate of potassa; the silver is procured in a pure form, chloride of potassium being produced, while carbonic acid and oxygen are evolved.

1182. II. Add lead to the alloy, and heat it on a porous cup of bone-ashes; the copper oxidates and fuses along with the lead; the fused oxides are absorbed by the porous cup, and the pure silver is left. The porous cup is usually termed a CUPEL, and the operation is termed CUPELLATION. Other metals which are not oxidated permanently by the united action of heat and air, are purified also in this manner.

1183. III. An acid liquid prepared by mixing 8 parts of sulphuric acid and 1 of nitre dissolves silver, but not copper.

1184. IV. Silver may be separated from its solutions in a metallic form by copper, zinc, and other metals, when it presents a crystallized appearance. Precipitated from the nitrate by mercury, it is obtained in an arborescent form, giving rise to what is commonly termed the ARBOR DIANÆ, or TREE OF DIANA; in this condition there is mercury associated with it.

1185. OXIDE OF SILVER.—Symb. ·Ag = silver 110 + oxygen 8.—Eq. 118.—Solid, insoluble in water, of a dark brownish-gray colour. PREPARED from the nitrate of silver by lime water, nitrate of lime being left in solution, and the oxide precipitated.

1186. Oxide of silver precipitated from the nitrate of silver by lime-water, and left in contact with ammonia, produces an ammoniuret which detonates with very great violence by heat or friction. Another detonating compound of silver will be described under Cyanogen.

1187. NITRATE OF SILVER, or LUNAR CAUSTIC.—Symb. ⁙N·Ag = nitric acid 54 + oxide of silver 118.—Eq. 172.—Solid, fusible, decomposed when heated above its fusing point. Soluble in water, crystallizes in plates. Corrodes animal and vegetable matter, and is much used as an escharotic. It is the basis of marking ink.

1188. PREPARED by the action of diluted nitric acid and metallic silver. Part of the acid is decomposed, oxidating the silver; the rest combines with the oxide.

1189. Chloride of Silver occurs native; and always presents itself as a curdy precipitate when muriatic acid or a solution containing chlorine is mixed with a solution containing silver. Symb. ClAg = chlorine 36 + silver 110.—Eq. 146.—White, solid, fused by heat, forming a horny looking mass when cold; it becomes of a dark violet colour when exposed to light. Insoluble in water, soluble in ammonia.

1190. Cyanide of Silver is formed when prussic acid is added to nitrate of silver.

1191. Sulphate of Silver is usually procured by mixing sulphate of soda with nitrate of silver.

1192. Carbonate of Silver is precipitated on adding an alkaline carbonate to the nitrate of silver.

Tests of Silver.

1193. A number of metals, as copper and zinc, precipitate metallic silver from its solutions; alkaline hydrosulphurets give a dark coloured precipitate, and chlorine or muriatic acid precipitates chloride of silver, which is soluble in ammonia. See 1189. Metallic silver may be easily procured from the chloride by heating it with an alkaline carbonate.

CHAP. III. GOLD.

1194. Symb. Au.—Eq. 200.—Sp. gr. 19.3.—Fuses at 2016° F. according to Daniell. Colour yellow, acquires a high lustre when polished; the most malleable and ductile of the metals. Dissolved by chlorine and bromine and acid compounds containing them in a free state, as the nitromuriatic acid, and nitrohydrobromic acid. Not tarnished, nor in any way affected by the action of air and moisture.

1195. Separated from copper and other oxidable metals by the operation of cupellation; and from some of its ores it is procured by the process of amalgamation; see silver. From silver it is separated by nitric or sulphuric acid.

1196. Oxide of Gold (protoxide).—Precipitated of a green colour when potassa is added to the chloride mixed with water.

1197. Peroxide of Gold.—Symb. ⁝Au.—Precipitated by potassa from the terchloride of gold. Many consider that it resembles acids more than oxides, and have termed it auric acid. Kept in contact with a strong solution of ammonia, a detonating compound is procured, the ammoniuret of gold.

1198. Terchloride of Gold.—Symb. 3Cl+Au = chlorine 108 + gold 200—Forms ruby-coloured crystals. Decomposed at a red heat. Soluble in water, alcohol, and sulphuric ether; the ethereal solution is sometimes used for gilding, depositing metallic gold on a number of substances, more especially when exposed to the light. Prepared by the action of nitromuriatic acid with gold, any excess of acid being expelled by heat. If it be heated to a temperature about 400°, part of its chlorine begins to escape, and chloride with terchloride of gold remains. Heated to redness, the chlorine is entirely dissipated, metallic gold being left.

Tests of Gold.

1199. Gold is precipitated from its solutions in the metallic form by many different substances, as green sulphate of iron, and a number of metals. Hydrosulphuric acid gives a dark coloured sulphuret. Borax heated with a small quantity of the solution acquires a brilliant ruby colour, and metallic gold is generally seen if the heat be long continued. Protomuriate of tin and metallic tin give a purple coloured precipitate, long known by the name of the PURPLE PRECIPITATE OF CASSIUS.

CHAP. IV. PLATINUM.

1200. Symb. Pt.—Eq. 96.—Sp. gr. 21.5. Resembles silver, but is not so white and has a less brilliant lustre; malleable and ductile. Not affected by air or moisture at any temperature; infusible in the most powerful furnaces, but melted in the flame of the oxyhydrogen blowpipe; at an elevated temperature it can be welded in the same manner as iron. Dissolved by chlorine and nitromuriatic acid. It is a bad conductor of heat; a coil of wire may be kept at a dull red heat by the slow combustion of coal-gas or spirit of wine, without any flame.

1201. In a minute state of division, it promotes the union of oxygen and hydrogen gases, absorbing them in large quantity, and becoming incandescent. See 439. PREPARED in this condition by adding muriate of ammonia to the bichloride of platinum in solution, and heating to redness the yellow coloured MURIATE OF AMMONIA AND PLATINUM which is thus precipitated; nothing is left but the metallic platinum in a minute state of division.

1202. OXIDE OF PLATINUM.—This term has been commonly applied to the compound which is more correctly termed the binoxide. It is precipitated when soda is added to a solution of the bichloride of platinum, and is obtained with difficulty in a pure form. Other oxides have been described, but they have not been applied to any important purpose.

1203. CHLORIDE OF PLATINUM is procured when the bichloride is subjected to a temperature of 450°. A red heat expels all the chlorine.

1204. BICHLORIDE OF PLATINUM is formed when this metal is digested in nitromuriatic acid, and the solution evaporated to dryness with a very gentle heat to expell excess of acid. It is soluble in water, and is the compound of platinum generally employed where the metal is required in solution. Bichloride of platinum is also soluble in alcohol and ether.

1205. AMMONIURET OF PLATINUM is a fulminating compound analogous to the ammoniuret of gold, and is prepared by combining the oxidated metal with ammonia.

1206. Platinum in solution tends to form many double salts, as chloride of potassium and platinum, which is precipitated when chlorine, potassium, and platinum, meet together in a concentrated solution.

Tests of Platinum.

1207. Muriate of ammonia gives a yellow precipitate, which leaves pure platinum when heated; iodide of potassium gives a very dark coloured precipitate, composed of iodine and platinum. Hydrosulphuric acid gives a dark brown precipitate, which becomes black when dry. By evaporating its solutions to redness, metallic platinum may be procured.

CHAP. V. PALLADIUM, &c.

1208. PALLADIUM, RHODIUM, OSMIUM, and IRIDIUM, are metals which have been procured from the ores of platinum, but they have not hitherto been procured in such quantity as to admit of their being employed extensively for any very important purpose.

1209. PLURANIUM is another metal which has been announced to exist in some of the ores of platinum, but farther experiments are required to confirm this statement.

ALLOYS AND AMALGAMS.

1210. Alloys are compounds of metals, and when mercury forms a part of the compound it is usually termed an Amalgam. These compounds are generally prepared by fusing together the different metals which are to be combined. The composition of some of the most important alloys, &c. is stated below.

1211. Brass = 3 parts of copper and 1 of zinc.

1212. Dutch Gold and Pinchbeck are composed of the same metals as brass, the proportions being varied.

1213. Bronze and Gun Metal = 8 or 9 parts of copper and 1 of tin.

1214. Bell Metal = 3 parts of copper and 1 of tin.

1215. Speculum Metal = 2 parts of copper and 1 of tin.

1216. Pewter is composed principally of tin with small portions of other metals.

1217. Plumber's Solder = 1 part of tin and 2 of lead.

1218. Tin-foil is composed principally of tin with a small portion of lead.

1219. Fusible Alloy = 8 parts of bismuth with 5 of lead and 3 of tin.

1220. Rose's Fusible Alloy = 2 parts of bismuth with 1 of lead and 1 of tin.

1221. Gold Coin = 11 parts of gold and 1 of copper.

1222. Standard Silver = 12 and 1–3d of silver with 1 of copper.

1223. Tinned Iron or Tin Plate consists of iron coated on the surface with tin.

1224. Many metals are often coated with gold or silver, by precipitating the gold or silver upon them.

1225. Amalgams containing gold and silver are often used for the same purpose, being spread upon the metals to be coated; the mercury is expelled afterwards by heat.

1226. Eight parts of the fusible alloy with 1 of mercury form an amalgam which is fluid at 170°.

1227. Zinc 1 part, with 1 of tin and 2 of mercury, form an amalgam much used for the electrical machine.

1228. Alloys of the alkaline metals are frequently prepared by applying heat to some of their compounds with the vegetable acids, after mixing with them the metal with which they are to be combined. Thus bismuth or antimony, heated with the bitartrate of potassa, gives an alloy of bismuth or antimony with potassium; the carbon of the tartaric acid decomposes the potassa.

PART III.

ORGANIC CHEMISTRY.

1229. ORGANIC CHEMISTRY includes the chemical history of all the products that are met with in the animal and vegetable kingdoms. In composition they differ widely from mineral substances, all of them being compounds, and, notwithstanding their number, consisting almost entirely of carbon, oxygen, hydrogen, and nitrogen. All of them contain carbon, and most of them hydrogen and oxygen; few vegetable matters have nitrogen, but it abounds in the animal kingdom.

1230. The particles in organic compounds appear to be grouped together in a more complicated manner than is observed in the mineral kingdom; they also contain associated with them minute and often variable proportions of potassa, soda, lime, magnesia, iron, and other mineral substances, formerly considered too trifling in quantity to affect materially their qualities, though recent researches have satisfactorily proved that even so minute a quantity as the 10,000dth part of some bodies, completely changes the character of the compound with which they may be combined. The remarks already made on isomeric compounds (see par. 68 *et seq.*) apply peculiarly to the products now under consideration.

1231. All the compounds of the animal and vegetable kingdoms are called PROXIMATE PRINCIPLES, their ULTIMATE PRINCIPLES being the elements of which they are composed.

1232. They are all decomposed by a red heat; if previously mixed with potassa, CARBONATE OF POTASSA is produced; and many of them, when subjected to a less elevated temperature with potassa, soda, or lime, produce OXALIC ACID, their carbon being less highly oxygenated.

1233. Heated in the open air, their carbon and hydrogen, uniting with its oxygen, form carbonic acid and water. The same products may be procured by heating them with oxide of copper, and from their amount the quantity of carbon and hydrogen is inferred in CONDUCTING THE ANALYSIS OF ORGANIC BODIES, 9 of water indicating 1 of hydrogen, and 22 of carbonic acid indicating 6 of carbon. The nitrogen being disengaged in a free state, its quantity is readily calculated when the carbonic acid has been removed by an alkali. Minute quantities of lime, potassa, soda, iron, &c. are ascertained by examining the ash that remains after the incineration of the organic matter. By deducting the weight of the ingredients previously estimated from the total amount of matter subjected to analysis, the amount of oxygen is determined.

1234. When heated, so as to exclude the action of the air, water and numerous inflammable compounds are disengaged, a carbonaceous residue being left, often very difficult of incineration. The products obtained in this manner are said to be formed by DESTRUCTIVE DISTILLATION.

1235. Most animal, and a few vegetable, substances give out ammonia when heated, in consequence of part of their nitrogen and hydrogen combining together.

1236. Organic compounds are remarkably prone to decomposition from the re-action of their elements, and in this circumstance differ widely from most inorganic bodies. The changes produced in this manner are included under the term FERMENTATION.

1237. The saccharine, vinous, and acetic fermentations, take place only under peculiar circumstances (see Sugar, Alcohol, and Acetic Acid).

1238. Most organic matters are subject to the PUTREFACTIVE FERMENTATION. Some organic principles, more especially oils and inflammable matters containing much carbon and hydrogen, are not decomposed in this manner.

1239. Air and moisture promote it. Many substances carefully excluded from these agents may be preserved for years. This is well illustrated by the mode in which some kinds of provisions are now preserved for long voyages.

1240. A moderate temperature promotes putrefaction. Heat retards it by expelling moisture, and cold prevents it altogether when the moisture is congealed.

1241. The products of putrefaction are water, bihydruret of carbon, carbonic acid, phosphureted and sulphureted hydrogen, acetic acid, ammonia in large quantities from most animal substances and vegetable matters containing nitrogen, and a dark coloured matter, containing a peculiar principle which agriculturists regard as the basis of the nutritive matter of soils, and which they term HUMINE. Other products are also disengaged, particularly odorous effluvia, and various miasmata, whose influence on the animal economy is well known, though they have hitherto eluded the most minute chemical analysis.

DIVISION I.—VEGETABLE CHEMISTRY.

GERMINATION AND VEGETATION.

1242. GERMINATION is the process by which a plant begins to grow from its seed. This process commences with the absorption of oxygen by the seed lobes or cotyledons, the production of carbonic acid gas, and the formation of saccharine matter, which is the first nourishment the plant receives as it exists in the seed. Water must be present to enable these changes to take place, and the most favourable temperature varies from 55° to 80°. Light retards germination.

1243. The GERM then increases, and the RADICLE descends into the earth to form the roots, while the PLUMULA ascends to constitute the stem.

1244. Germination cannot take place at a great depth, in consequence of the absence of oxygen. Chlorine, by facilitating the evolution of oxygen from water, promotes germination.

1245. VEGETATION includes all those processes by which the growth of the plant is carried on after germination. They are analogous to those observed in the animal economy.

1246. By ABSORPTION from the soil, the roots of plants take up a large share of nutritious matter; they also receive a considerable proportion of nourishment through the medium of their leaves, and accordingly they can, in many cases, be supported in this manner alone. In the latter case the carbon seems to be derived principally, if not entirely, from the carbonic acid of the atmosphere.

1247. By CIRCULATION the juices of plants are brought to the leaves, and there, by a process analogous to RESPIRATION, undergo those changes by which they are more particularly fitted to nourish the plant.

1248. By SECRETION the different proximate principles are elaborated from the sap.

1249. Light has a very important influence, not only on the colour of plants but also on the process of respiration, favouring very much the decomposition of carbonic acid, the absorption of carbon, and the elimination of oxygen.

1250. In the present state of vegetable chemistry, the products of vegetation may be considered most conveniently under the following heads:—

1. Vegetable Acids.
2. Vegetable Alkalis.
3. Oily and Resinous Substances.
4. Starch, Sugar, Gum, &c.
5. Colouring Matter.
6. Asparagin, Ulmin, &c.
7. Alcohol, and products obtained from it. Saccharine and Vinous Fermentation.

ORDER I. VEGETABLE ACIDS.

CHAP. I. OXALIC ACID.

1251. Symb. $\vdots C^2$ = oxygen 24 + carbon 12.—Eq. 36.—Solid, crystallized, combined with water. Symb. of Cr. oxalic acid, $3\cdot H + \vdots C^2$. Crystals prismatic and flattened; fuse at 209°, at 212° they lose 2 eqs. of water; heated suddenly they are decomposed. Soluble in about 15 cold, and in their own weight of boiling water. Heated with sulphuric acid, the water of crystallization is removed, carbonic oxide and carbonic acid are disengaged. Taste extremely sour. Reddens powerfully the vegetable blues; is precipitated by lime-water. Much employed as a test for lime, the oxalate of lime being very insoluble.

1252. Oxalic acid is a powerful poison in the quantity of 2 or 3 drachms, and is frequently mistaken for epsom salts. They may be distinguished easily in the following manner:—

Tests.	Oxalic Acid.	Sulphate of Magnesia.
Taste.	Acid.	Bitter.
Thrown on red-hot cinders.	Completely dissipated.	A white mass remains.
Vegetable blue.	Reddened.	Not affected.
Alkaline carbonates.	Effervesce.	Give a white precipitate.

1253. ANTIDOTES.—Chalk and magnesia. The oxalates of lime or magnesia which are formed being inert.

1254. PREPARATION.—I. By digesting sugar or waste syrups with nitric acid, a syrupy fluid is formed; the crystals deposited are purified by solution and crystallization.

1255. II. By heating potassa with saw-dust, or some other vegetable matters, to a temperature between 400° and 450°, the potassa being afterwards removed by a stronger acid.

1256. III. From the juice of the OXALIS ACETOSELLA, and some other plants, in which it is found in combination with potassa.

1257. Oxalates are usually prepared by neutralizing oxalic acid with the salifiable bases.

1258. OXAMIDE is the name given to a compound of 12 of carbon + 14 nitrogen + 16 oxygen + 2 hydrogen, and it is formed by destructive distillation from oxalate of ammonia.

CHAP. II. ACETIC ACID.

1259. Symb. A = ⋮H^3C^4 or 27 water + 24 carbon.—Eq. 51.—Liquid and combined with 1 eq. water in its most concentrated form. Crystallizes at 32°. Blisters the skin. Very volatile; odour pungent and fragrant, vapour inflammable. Sp. gr. of liquid acid 1.0629. The specific gravity alone does not indicate its strength, as this increases and then diminishes on adding successive quantities of water. By neutralizing with carbonate of potassa, or carbonate of lime, its precise value is ascertained, 70 of the first or 50 of the second, indicating 51 of acid.

1260. PREPARATION.—Vinegar contains acetic acid; by distillation it is separated from the vegetable matter in solution, but much water rises along with it. On neutralizing the distilled acid with potassa, soda, or lime, the water may be expelled by evaporation. Sulphuric acid, added to the compound thus formed, unites with the base, concentrated acetic acid being separated by distillation. If the acetic acid be combined with oxide of copper, it may be separated by heat alone; a portion of it is then decomposed, and a peculiar fluid resembling alcohol in some of its properties is formed—the PYRO-ACETIC SPIRIT.

1261. The acid fluid procured by the destructive distillation of wood may be made to yield acetic acid by a similar process. The impure acid procured at first from wood is usually termed PYROLIGNEOUS ACID.

1262. Alcohol may also be converted into acetic acid by exposing a small quantity for some time to the action of air and spongy platinum in a large glass vessel.

1263. Vinegar is usually prepared by subjecting liquids that have undergone the vinous fermentation to the action of the air; much oxygen is then absorbed. Many solutions of vegetable matter, which have not previously passed through the vinous fermentation, produce acetic acid during fermentation.

CHAP. III. TARTARIC ACID.

(RACEMIC AND PARATARTARIC ACIDS.)

1264. Symb. T. = oxygen 40 + carbon 24 + hydrogen 2.—Eq. 66.—Crystallizes in prisms = 66 acid + 9 of water; eq. 75; soluble in alcohol and water. It has a great tendency to form double salts. Added in excess to potassa, cream of tartar is precipitated, a character by which it is distinguished from other acids. By digestion in nitric acid it is converted into oxalic acid. Exposed to heat it is decomposed; by destructive distillation a new acid is obtained, termed the PYRO-TARTARIC ACID.

1265. PREPARED from bitartrate of potassa (cream of tartar); 50 parts of carbonate of lime (chalk) mixed with 200 of common cream of tartar, and thrown into boiling water, give a precipitate of 94 of tartrate of lime, carbonic acid being discharged with effervescence, and tartrate of potassa remaining in solution. By adding muriate of lime to the solution, 94 parts more of tartrate of lime are precipitated, and muriate of potassa is left in solution. On adding 40 of real sulphuric acid to 94 of tartrate of lime, 68 of sulphate of lime are produced, and 66 of tartaric acid separated, which may be removed by water from the greater portion of the sulphate of lime. The solution of tartaric acid being evaporated, a little sulphate of lime is first separated, and crystals of tartaric acid afterwards deposited, which are purified by repeated solution and crystallization.

1266. RACEMIC and PARATARTARIC ACIDS are the names which have been given to a vegetable acid isomeric with tartaric acid, which differs from it in several circumstances of minor importance. It is usually obtained from the juice of some varieties of grape. Dr Thomson considers it a more powerful açid than tartaric acid, and that 2 eqs. of water are combined with it when crystallized.

CHAP. IV. CITRIC, MALIC, BENZOIC, GALLIC, SUCCINIC, &c. ACIDS.

1267. CITRIC ACID.—Symb. *Ct.* = oxygen 32 + carbon 24 + hydrogen 2.—Eq. 58.—Combined with 1 or more eqs. of water according to the manner in which it is crystallized. Much employed when dissolved in water as a substitute for lemon juice, as its crystals can be preserved for any length of time. Decomposed by heat. It has a great resemblance to tartaric acid in its general characters, but does not form a sparingly soluble compound when added in excess to potassa.

1268. Prepared by adding the juice of limes or lemons to chalk, and decomposing the citrate of lime which is procured in this manner, according to the process adopted in the preparation of tartaric acid.

1269. MALIC ACID, called also SORBIC ACID, is found in the juice of the berries of the Pyrus aucuparia, of the apple, and in a number of other vegetable productions. It has been of late regarded as isomeric with the citric acid.

1270. BENZOIC ACID.—Symb. *B.*—Eq. 120.—But very different estimates have been given lately of its composition and chemical equivalent. Liebig and Wohler regard it as a compound of a peculiar principle, which they have termed BENZULE, and oxygen. Its taste is sweet and aromatic. Heat melts it, and then converts it into vapour; it condenses in crystals. By an elevated temperature it is decomposed, taking fire if in contact with air. It is very sparingly soluble in water. Employed principally on account of its fragrant odour, and to precipitate peroxide of iron from its neutral solutions, being used in the form of BENZOATE OF AMMONIA.

1271. PREPARED by sublimation from GUM BENZOIN, mixed previously with sand; or by boiling the gum in water with carbonate of soda, when benzoate of soda is procured in solution, the benzoic acid being then precipitated from the solution by sulphuric acid. It is made to assume the usual form by sublimation.

1272. GALLIC ACID is obtained principally from gall-nuts, by exposing an infusion of them for some weeks to the air. It is soluble in water and alcohol, does not precipitate gelatine, and is particularly distinguished by the dark colour which it produces with persalts of iron, forming with the peroxide the black compound which is the basis of ink. ELLAGIC ACID constitutes the insoluble matter that is found at the bottom of the infusion; and PYROGALLIC ACID is a peculiar acid which is formed when the gallic acid is heated. By subjecting it to a particular temperature, another compound is formed, which has been termed METAGALLIC ACID.

1273. SUCCINIC ACID is prepared by distillation from amber, condensing in the neck of the retort. By adding nitric acid and evaporating to dryness, it is procured in a pure form. It is employed principally for precipitating peroxide of iron, and is used in the form of succinate of ammonia.

1274. MUCIC or SACCHOLACTIC ACID is formed when gum or sugar of milk is digested in nitric acid. Guerin considers the acid produced by gum and nitric acid to be peculiar, and has termed it OXALHYDRIC ACID. From its composition it may be regarded as a compound of oxalic acid and hydrogen.

1275. CAMPHORIC ACID is procured by the action of nitric acid on camphor.

1276. MOROXYLIC ACID is extracted from the bark of the white mulberry tree.

1277. BOLETIC ACID is found in the Boletus pseudo-igniarius.

1278. IGASURIC ACID; see Strychnia. MECONIC ACID; see Morphia. KINIC ACID; see Quina.

1279. PECTIC ACID (πηκτις, coagulum) is found in most plants, and derives its name from its tendency to form a gelatinous mass with water.

1280. LACTIC ACID is found in the juice of the beet-root, in sour milk, and in most animal fluids.

1281. Numerous other acids have been described lately, all of which are comparatively of very secondary importance. Of these it will be sufficient to mention the names in this place.

1282. VALERIANIC ACID, procured from Valeriana officinalis.

1283. CARBAZOTIC and INDIGOTIC ACIDS, formed by the action of nitric acid on indigo.

1284. ROCELLIC ACID is prepared from the Rocella tinctoria.

1285. LACTUCIC ACID is obtained from the Lactuca virosa.

1286. SUBERIC ACID is formed by the action of nitric acid on cork.

CHAP. V. PRUSSIC OR HYDROCYANIC ACID, CYANOGEN, AND THEIR COMPOUNDS.

1287. This subject has become so extensive, and at the same time so complicated, that I have placed together in the present chapter all the most important facts connected with it. Though hydrocyanic acid may be obtained from some plants, it is prepared principally from animal matters, which, when heated with potassa and iron, form the cyanide of potassium and iron, a compound that, with water, gives ferroprussiate of potassa, from which prussic acid, cyanogen, and the greater number of its compounds are commonly procured.—See Hydrocyanic Acid.

SECT. I. *Cyanogen, or Bicarburet of Nitrogen.*

1288. Symb. C^2N = carbon 12 + nitrogen 14.—Eq. 26 by W, □ by V.—Sp. gr. 1.8.—W. of 100 c. i. 55 grs.—Condensed into a liquid at 45°, by a pressure of 3.6 atmospheres. Gaseous, transparent, colourless, inflammable, burns with a beautiful purple flame, producing carbonic acid and nitrogen.

Fig. 59.

Materials.			Products.
Cyanogen 1 eq.	Nitrogen	14	14 Nitrogen.
	Carbon	6	22 Carbonic Acid.
	Carbon	6	
Oxygen 4 eqs.	Oxygen	16	
	Oxygen	16	22 Carbonic Acid.

Symb. $C^2N\&O^4 = 2{:}C\&N.$

Odour peculiar, pungent, and penetrating. Absorbed by water and alcohol. Its compounds when not acid are termed CYANURETS or CYANIDES; the former is preferable, as cyanogen resembles inflammables more than oxygen, but the latter is perhaps in most general use in this country. Termed Cyanogen from κυανος, purple, being one of the ingredients of prussian blue.

1289. PREPARED from the bicyanide of mercury by exposing it to heat, collecting the gas over mercury, as water condenses it. Part of the mercury is volatilized, and a black porous matter remains, regarded formerly as charcoal from decomposed cyanogen, but now considered as a peculiar compound, containing the same elements as cyanogen, united in the same proportions. Professor Johnston has lately shown that it is formed in a great variety of cases from cyanogen and its compounds. It has been termed PARACYANOGEN.

SECT. II. *Hydrocyanic or Prussic Acid.*

1290. Symb. HC^2N or P = cyanogen 26 + hydrogen 1.—Eq. 27 by W, □□ by V.—Sp. gr. of liquid acid at 45°, 0.705. Volatile, boils at 80°, freezes at zero. Transparent, limpid, colourless; odour strong, pungent, penetrating, inducing headach, nausea, and even fainting, when much of its vapour is inspired. When diluted, it has the odour and taste of the peach-blossom and bitter almond. The taste of the strong acid is cool at first, but a sensation of warmth is soon perceived. Being one of the most virulent poisons, and perhaps more rapidly fatal in its action than any other poison, much care is required in experimenting with it. A single drop introduced into the throat of a large dog, kills it, after a few hurried inspirations. Inflammable; water, carbonic acid, and nitrogen, are the products of its combustion. Symb. $HC^2N\&O^5$ = 2:C&·H&N.

1291. Feebly acid, does not neutralize alkalis, has little action with litmus paper, and does not discharge carbonic acid from carbonates. Begins to be decomposed a few hours after it has been prepared, even when excluded from the light; much ammonia being formed, and a black carbonaceous matter left. Diluted with water, it may be kept without decomposition for years. Resolved by galvanism into cyanogen and hydrogen. Potassium heated in its vapour forms cyanide of potassium and leaves hydrogen. Decomposed rapidly by chlorine, which is used as an antidote to it.

1292. *Method of estimating the Strength of Prussic Acid.*—The specific gravity of the acid, when combined with different proportions of water, not varying sufficiently to admit of its strength being ascertained with facility in this manner, numerous other methods have been proposed. Dr Ure recommended the absolute quantity of prussic acid in any solution to be determined by the quantity of finely pounded peroxide of mercury which it can dissolve, when shaken with it in a phial. Lately, the nitrate of silver has been preferred, cyanide of silver being precipitated, and nitric acid with water left. The water is produced when the oxygen of the oxide unites with the hydrogen of the acid, leaving the cyanogen and the metal in combination. The cyanide of silver appears as a dense curdy precipitate, very similar in appearance at first to the chloride of silver, but easily distinguished from it, as it does not darken on exposure to light. When heated, the cyanogen inflames, and the metallic silver is left. Every 110 grs. of silver indicate 26 of cyanogen, and consequently 27 of prussic acid.

1293. *Detection of Sulphuric and Muriatic Acids.*—Sulphuric acid may be detected in the usual manner by a solution of baryta, or of any of its salts. The presence of muriatic acid is ascertained most easily by adding to a small portion of the acid a little ammonia, and exposing it to a very gentle heat. If pure, nothing remains in a short time, hydrocyanate of ammonia alone having been formed, which is entirely dissipated. But when muriatic acid is present, muriate of ammonia remains, which may be dissolved in water, chloride of silver being then precipitated by the addition of the nitrate, which darkens on exposure to light.

1294. PREPARATION. I. *By passing hydrosulphuric acid over heated bicyanide of mercury, till it is rendered black, and expelling the concentrated acid by a gentle heat.*—2 eqs. of hydrosulphuric acid = 34, decompose 1 of the bicyanide of mercury = 252, producing 2 eqs. of hydrocyanic acid and 1 of bisulphuret of mercury.

Fig. 58.

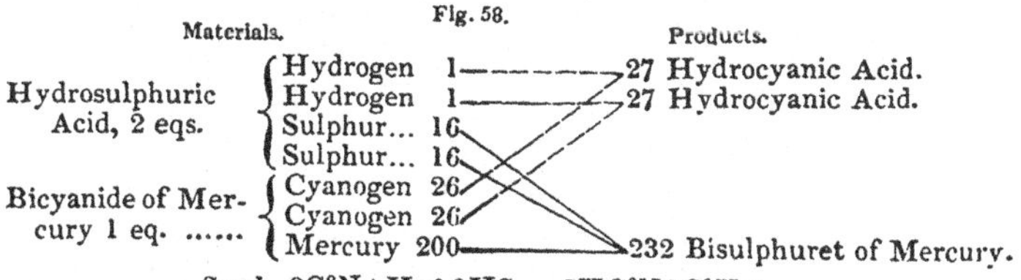

Symb. $2C^2N+Hg\&2HS = 2HC^2N\&S^2Hg$.

1295. II. *By passing the hydrosulphuric acid through the bicyanide in solution, dilute prussic acid is procured, the bisulphuret being precipitated.*—Carbonate of lead must be added so long as it is blackened, to remove any excess of sulphureted hydrogen.

1296. III. *By distillation from bicyanide of mercury and muriatic acid.*—2 eqs. of muriatic acid decompose 1 of bicyanide of mercury, 2 of hydrocyanic acid and 1 of bichloride of mercury are formed, the latter being left in the retort. When the acid is required highly concentrated, it must be passed over chloride of calcium and carbonate of lime as prepared by this process, the carbonate retaining any muriatic acid, and the chloride taking any water that may have been associated with it. Symb. $2C^2N+Hg\&2HCl = 2HC^2N\&Cl^2Hg$.

1297. The Bicyanide of Mercury is prepared most conveniently by adding 2 eqs. of hydrocyanic acid to 1 of peroxide of mercury in fine powder; the 2 eqs. of oxygen in the peroxide uniting with the 2 eqs. of hydrogen in the acid produce 2 of water, and 1 eq. of bicyanide of mercury is deposited in crystals. This process I was led to propose some time ago, having found it a much more convenient method of preparing the bicyanide than that generally resorted to. The hydrocyanic acid employed may be prepared from the ferroprussiate of potassa by sulphuric acid.

Fig. 59.

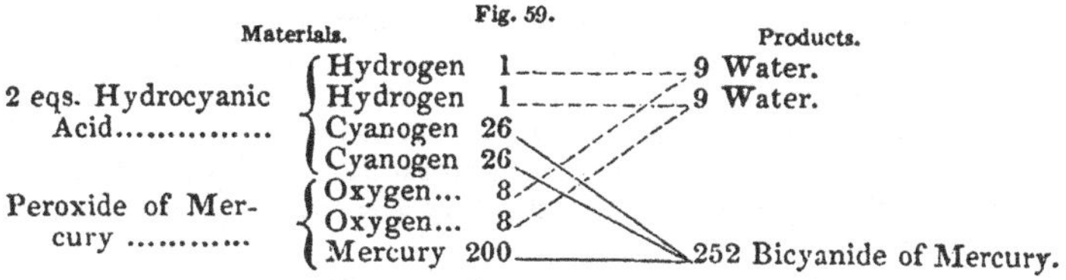

1298. The process at present in general use for the preparation of the bicyanide consists in precipitating perferrocyanate of iron (prussian blue) from the ferrocyanate of potassa by a salt of iron, transferring the prussic acid it contains to peroxide of mercury by boiling them in water, when a compound of 2 eqs. of acid and 1 of peroxide is procured, which gives 1 of crystallized bicyanide of mercury, and 2 eqs. of water on concentrating the solution.

1299. IV. *By the action of sulphuric acid diluted with water on the ferrocyanate of potassa.*—Ferrocyanate of potassa may be resolved into potassa, hydrocyanic acid, and cyanide of iron. Its component parts are 2 eqs. of potassa and 1 of ferrocyanic acid, which is composed of 2 eqs. of hydrocyanic acid and 1 of cyanide of iron. By the action of 2 eqs. of sulphuric acid, 2 of sulphate of potassa are formed, with which the cyanide of iron is left, the prussic acid being separated by distillation.

Fig. 60.

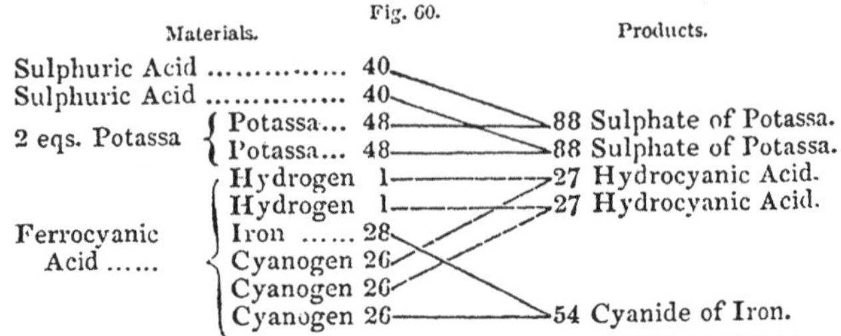

Symb. $3C^{2}N+Fe+2H+2\cdot K\&2\vdots S = 2\vdots S\cdot K\&C^{2}NFe\&2HC^{2}N.$

1300. Professor Everitt has lately given a different view of the changes that attend the production of hydrocyanic acid from this salt. He recommends 212.5 grains of the crystals to be dissolved in 2 fluid ounces of water, adding 600 grains of diluted sulphuric acid, containing 20 per cent. of real acid (120 of real acid). The receiver must be kept very cold, and a portion of water may be put into it; 41 grains of hydrocyanic acid pass off with the first third of the water.

1301. According to his views, the ferrocyanate has *three* eqs. of combined water, and with the preceding proportions, 2 eqs. of the salt are decomposed by 6 of sulphuric acid. Three of bisulphate of potassa are formed, three of hydrocyanic acid are separated by distillation, three of water are set at liberty, and a new yellow salt formed, composed of 1 eq. of cyanide of potassium and 2 of cyanide of iron.

$$6\vdots S\&3\cdot H+2(2C^{2}NK+C^{2}NFe) = 3(2\vdots S+\cdot K)\&3HC^{2}N\&(C^{2}NK+2C^{2}NFe).$$

The yellow coloured salt ultimately produces prussian blue, but the manner in which this change takes place has not been minutely studied.

1302. From the acid procured in this manner, concentrated hydrocyanic acid is easily procured by the action of dry chloride of calcium in fine powder. It unites with the water, and the absolute prussic acid floats upon the solution thus formed. The chloride must be added in small quantities at a time, otherwise heat will be evolved so rapidly during its action on the water, that much of the acid will be converted into vapour with ebullition, to the great danger of the operator. The phial in which the experiment is made should be long and narrow, and surrounded with a freezing mixture.

1303. V. *By decomposing hydrocyanate of potassa by tartaric acid.*—2 eqs. of tartaric acid added to 1 eq. of hydrocyanate of potassa, precipitate 1 eq. of bitartrate of potassa and leave the prussic acid in solution. A small quantity of bitartrate also remains in solution.

Fig. 61.

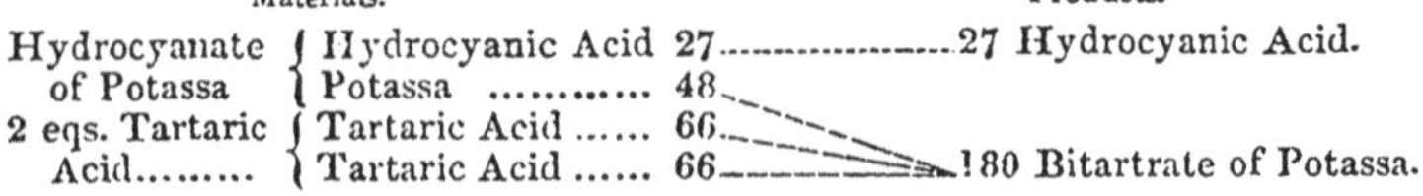

Symb. $HC^{2}N+\cdot K\&2T = HC^{2}N\&2T+\cdot K.$

The hydrocyanate of potassa is procured by decomposing the cyanide of potassium by water, the cyanogen taking the hydrogen, while the potassium combines with the oxygen.

Fig. 62.

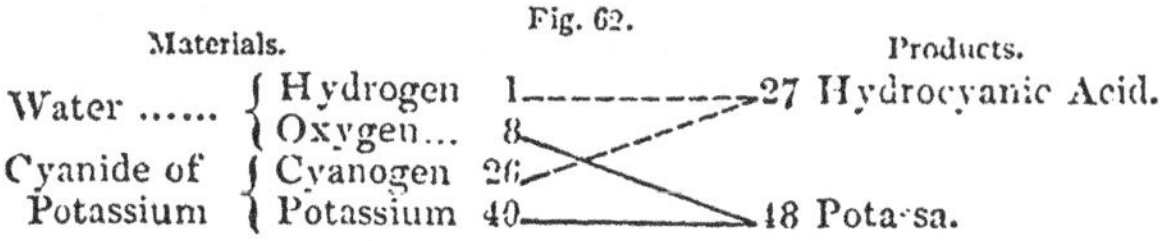

Symb. $C^2N+K\&\cdot H = HC^2N\&\cdot K = HC^2N+\cdot K$.

The cyanide of potassium is procured by heating the ferrocyanate of potassa to expel water, and then subjecting what remains to a red heat. The first products are cyanide of potassium and cyanide of iron; when these are carefully subjected to a red heat, the cyanide of iron is decomposed, and the cyanide of potassium left, which is separated by water, and converted into hydrocyanate of potassa.

Fig. 63.

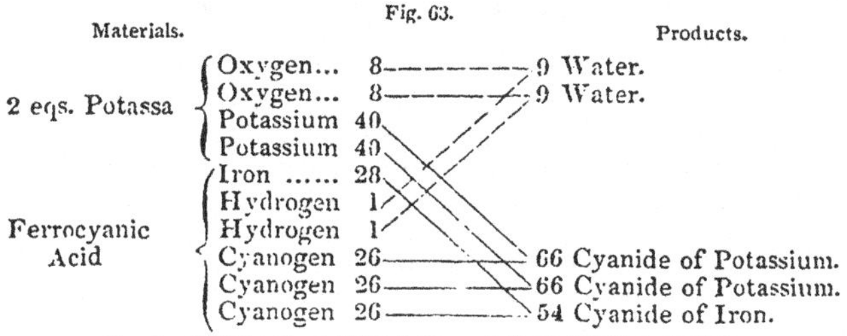

Symb. $3C^2N+Fe+2H+2\cdot K = 2\cdot H\&2C^2NK\&C^2NFe$.

1304. The acid known by the name of Scheele's Acid, contains very nearly 1 grain of pure hydrocyanic in 16 grains.

Tests of Hydrocyanic Acid.

1305. Oxide of Iron mixed with peroxide of iron is the best test of prussic acid, producing with it the deep blue pigment, prussian blue. The oxides act most efficiently when newly precipitated from any of their soluble salts. The following is the method of proceeding :—

I. Distil the suspected fluid with a water bath, acidulating previously with acetic acid.

II. Add a portion to a drop or two of sulphate and persulphate of iron in solution. No apparent effect is produced.

III. Add potassa to unite with the sulphuric acid, and precipitate oxides of iron. Part of the hydrocyanic acid decomposes the oxide, forming water and cyanide of iron. This cyanide uniting with 2 eqs. of undecomposed hydrocyanic acid produces ferrocyanic acid, which forms the prussian blue with the peroxide. The oxide is necessary to form the ferrocyanic acid, and the peroxide is equally necessary to form the prussian blue when this acid has been produced.

IV. If a blue colour be not observed in the precipitate, add slowly muriatic acid, as it may perhaps be disguised by an excess of oxides; if no blue be seen as the oxides in excess are dissolved, the suspected fluid is free from hydrocyanic acid.

1306. Oxide of Copper has also been proposed as a test of hydrocyanic acid. Precipitated from the sulphate by potassa, its oxygen takes hydrogen from the hydrocyanic acid, while the cyanogen and copper remain, forming a white curdy precipitate. Dilute muriatic acid removes any excess of oxide of copper.

1307. Nitrate of Silver has also been used as a test of this acid, and is also now employed in estimating its amount in liquids known to contain it; it is so delicate in its action, that it can detect hydrocyanic acid in 10,000 parts of water. Mr Barry recommends the acid to be mixed with a little acetic acid before adding the nitrate of silver.

Sect. III. *Cyanic and Fulminic Acids.*

1308. Symb. $\cdot C^2N$.—Very different statements have been made of late years as to the precise constitution of these compounds, but it is now considered that there are 2 acids, composed of 1 eq. of oxygen and 1 of cyanogen, isomeric with each other, and to these the provisional names of Cyanic and Fulminic Acids have been given.

1309. Cyanic Acid is formed when cyanogen acts on water containing an alkali, the water being decomposed, while cyanic and hydrocyanic acids are both produced, as the cyanogen combines with the elements of the water. It may also be procured in combination with potassa by heating cyanide of potassium and iron (prepared by heating ferrocyanate of potassa and expelling water) with peroxide of manganese, the cyanogen and potassium being oxidated by the peroxide. Alcohol dissolves the cyanate, and deposits it in tabular crystals. This acid is not applied to any important purpose. It is decomposed with extreme facility, and resolved by the action of water into bicarbonate of ammonia.

1310. Fulminic Acid is prepared from fulminate of mercury, according to the process recommended by Professor Ed. Davy. One part of the fulminate is mixed with 2 of zinc and 18 of water. Fulminate of zinc is thus procured, from which the oxide of zinc is removed by baryta, and sulphuric acid separates the baryta. It is liquid, volatile, poisonous, has an odour like prussic acid, and is easily decomposed.

Fulminates of Mercury and Silver are the only important compounds of this acid. They are prepared by the action of 100 parts of the metal, 950 of diluted nitric acid, sp. gr. 1.35, and 850 of alcohol, sp. gr. 0.835. A pernitrate of mercury, or nitrate of silver, having been formed by the action of the acid and the metal, the alcohol is added, when a portion of its carbon uniting with oxygen and nitrogen produces fulminic acid, which is deposited in combination with the metallic oxide, forming the fulminating compounds from which the acid has been named. Heat, friction with sand, or any other hard substance, and the action of acids, cause the fulminating compounds to explode powerfully; the fulminating silver is much more violent in its action than fulminating mercury, and much more easily detonated.

Sect. IV. *Ferrocyanic Acid, Ferrocyanate of Potassa, and Perferrocyanate of Iron.*

1311. Ferrocyanic Acid.—Symb. $6C+^3N+Fe+2H$, or C^2NFe+P^2.—Many chemists do not concur with those who have termed this compound an acid; the common term, however, has been adopted here, as it is at present so much in use. It is often called ferroprussic or hydroferrocyanic acid.

Crystallizes in yellow cubical crystals, reddens vegetable blues, and acts more powerfully as an acid than the hydrocyanic. Not volatile, nor poisonous in small doses; decomposed by light. It is separated from its combinations by galvanism, appearing at the positive pole of the galvanic battery.

1312. Prepared from the ferroprussiate of potassa by tartaric acid, which precipitates the potassa as a bitartrate. Another process consists in mixing ferrocyanate of potassa with hydrosulphuret of baryta in solution; ferrocyanate of baryta is precipitated, from which the baryta is separated by sulphuric acid.

1313. FERROCYANATE OF POTASSA.—Symb. $C^2NFe+2HC^2N+2{\cdot}K$ = ferrocyanic acid 108 + potassa 96.—Eq. 204.—Solid, gives 4-sided tabular crystals, lemon-yellow colour, soluble in less than their weight of boiling water. It is often represented as a compound of 2 eqs. of water, 2 of cyanide of potassium, and 1 of cyanide of iron = $2{\cdot}H+2C^2NK+C^2NFe$.

1314. RED FERROCYANATE OF POTASSA is the name given to a compound prepared by transmitting chlorine through a solution of ferrocyanate of potassa; a complicated reaction takes place. It is particularly distinguished by precipitating Prussian blue with salts of the protoxide of iron.

1315. Ferrocyanate of potassa is prepared by heating animal matters with carbonate of potassa and iron; the solid formed is lixiviated, and from the solution thus obtained crystals are procured by evaporation.

Much employed in the preparation of prussic acid, and as a test for metals in solution.

1316. PERFERROCYANATE OF IRON, commonly called PRUSSIAN BLUE. Inflammable, employed principally as a pigment. Precipitated when a per-salt of iron is added to the ferrocyanate of potassa in solution; or formed by the white precipitate from a proto-salt absorbing oxygen from the air; it is in this manner that Prussian blue is procured from the pale blue (see par. 966) precipitate which green vitriol gives with the ferrocyanate. Usually considered a compound of ferrocyanic acid and peroxide of iron, but very different opinions are entertained with respect to the manner in which its elements are united.

SECT. V. *Sulphocyanic Acid, &c.*

1317. SULPHOCYANIC ACID called also HYDROSULPHOCYANIC ACID.—Symb. HC^2N+S^2 = hydrocyanic acid 27 + sulphur 32.—Eq. 59.—Crystallizes from its aqueous solution in 6-sided prisms. Particularly distinguished by the deep red colour which it produces with the per-salts of iron.

1318. PREPARED by distillation from the sulphocyanate of potassa, sulphuric acid being added to retain the potassa. Sulphocyanate of potassa is prepared by heating ferrocyanate of potassa to redness in a crucible with sulphur; sulphocyanuret of potassium is thus produced; by pouring it into water, part of the water is decomposed, the sulphur and cyanogen combining with the hydrogen to form the acid, while the oxygen unites with the potassium to form potassa.

1319. BISULPHURET OF CYANOGEN.—Symb. S^2C^2N = sulphur 32 + cyanogen 26.—Eq. 58.—The base of sulphocyanic acid, prepared, though not at first in a state of purity, by the action of chlorine with heated sulphocyanuret of potassium. Different compounds of cyanogen with less sulphur have been noticed.

1320. The five following compounds have been discovered lately by Liebig :—

I. MELLON is the name given to the yellow coloured powder which remains after heating dry bisulphuret of cyanogen. By a high temperature it is converted into nitrogen and cyanogen gases.

II. MELAM consists of carbon, hydrogen, and nitrogen; it is also a yellow powder, and is formed by heating sulphocyanate of ammonia.

III. MELAMINE is composed of 6 eqs. of carbon, 6 of nitrogen, and 6 of hydrogen. It is produced when mellon or melam is boiled with a solution of potassa, being deposited in colourless crystals.

IV. AMMELINE remains in solution in the preceding process as the melamine is deposited; acetic acid unites with the potassa and precipitates it.

V. AMMELIDE is formed by the action of strong sulphuric acid on melam or melamine; alcohol added to the solution precipitates the ammelide.

1321. Many other combinations of cyanogen have attracted attention of late, but they are comparatively of much less consequence than the preceding compounds.

ORDER II. VEGETABLE ALKALIS.

1322. The term VEGETABLE ALKALI or ALKALOID is applied to all those proximate vegetable principles which have alkaline properties. Serturner, Duncan, Robiquet, Pelletier and Caventou, are the individuals who have made the most important discoveries in developing this branch of organic chemistry.

1323. All the vegetable alkalis occur in combination with acids; all of them, according to recent observations, contain nitrogen, and their alkalinity has been attributed to the presence of ammonia, formed by the union of their nitrogen with a portion of hydrogen; an opinion rendered more probable by their evolving ammonia, in numerous processes during which they are decomposed, as when narcotine is subjected to the action of a weak galvanic pile.

1324. Iodine combines with most of the vegetable alkalis in atomic proportions. Hydriodic and iodic acids generally act in the same manner. MORPHIA, however, is peculiar in its action with iodine, being quickly decomposed and giving hydrogen to part of the iodine, while a new compound remains, which unites with the rest. When iodic acid acts with morphia, it is decomposed, and a red matter is produced, similar in appearance to what is formed with nitric acid and this vegetable alkali.

1325. The alkaloids are composed of carbon, oxygen, hydrogen, and nitrogen; decomposed by heat with the evolution of ammonia, when the air is excluded; generally insoluble in water, soluble in hot alcohol; extremely energetic in their action on the animal economy. Some affect the vegetable colours very feebly, and narcotine does not act on them at all: but most of them affect colouring matter in the same way as the common alkalis. With the acids they form neutral salts, in which the characteristic properties of the alkaloid on the animal economy is generally observed. At 212°, they expel ammonia from its salts; most of them precipitate iodic acid.

1326. They are usually procured by extracting the vegetable salt in which they are found, separating the acid with which they are combined by an alkali or earth, and purifying them by repeated solution and crystallization.

CHAP. I. MORPHIA, AND THE OTHER VEGETABLE ALKALIS IN OPIUM.—MECONIC ACID.

1327. MORPHIA.—Eq. 290 = carbon 210 + oxygen 48 + hydrogen 18 + nitrogen 14. This is one of the most important of the vegetable alkalis, and the principle on which the narcotic properties of opium depend; insoluble in water; dissolved by boiling alcohol, crystallizing in 6-sided prisms as the solution cools; tasteless when pure. Extremely bitter when rendered soluble by alcohol or an acid. Neutralizes acids. Its alcoholic solution affects test-paper in the same manner as the common alkalis. Decomposes nitric acid, separating nitrous acid, and producing a deep red colour, which soon fades, as the nitrous acid escapes.

Decomposes iodic acid, and separates iodine, by which it is distinguished from the other vegetable alkalis.

1328. PREPARATION. Macerate opium in water; filter; the solution contains meconate of morphia. Boil the solution with magnesia; the morphia is precipitated, the meconic acid combining with the magnesia. On boiling the precipitated morphia in alcohol with animal charcoal, and filtering the hot solution obtained in this manner, crystals of morphia are deposited as it cools. The charcoal destroys adhering colouring matter.

Potassa, soda, ammonia, or lime, may be substituted for the magnesia in separating meconic acid.

1329. The salts of morphia may be procured by neutralizing morphia with the different acids diluted with water. The MURIATE OF MORPHIA, much employed in medicine, and introduced by Dr William Gregory into this country, is prepared most conveniently by the following process, as modified by Dr Gregory and by Dr Robertson. Muriate of lime is added to meconate of morphia in solution, prepared by the action of water on opium. Meconate of lime is precipitated; muriate of morphia remains in solution. It is purified by repeated crystallization, and pressing between folds of bibulous paper, to remove soluble impurities. Animal charcoal boiled with the solution removes much of the colouring matter with which the morphia is associated in the opium. About 10 parts of opium are required to give 1 of the muriate of morphia.

The muriate of morphia is obtained in small tufts of acicular crystals, a form very frequently observed in the crystallization of many vegetable salts. It is very soluble in water. It is frequently associated with muriate of codeia and muriate of narcotine. The narcotine may be detected by adding potassa in excess to the solution of the salt; morphia is precipitated and then dissolved by the excess of alkali, but the narcotine remains undissolved.

1330. NARCOTINE is another vegetable alkali which was removed from the list of vegetable alkalis, and again restored, its alkaline properties being now generally admitted; they are extremely feeble. It is soluble in alcohol, ether, and oils, but insoluble in water.

The unpleasant effects which so often attend the exhibition of opium, as headach, &c. are now very generally attributed, at least in part, to the narcotine.

Extracted by digesting opium in sulphuric ether, which does not dissolve morphia or meconate of morphia. The ether deposits narcotine on evaporation.

1331. ACETATE OF MORPHIA may be prepared by a process similar to that for the preparation of the muriate, or by dissolving morphia in acetic acid.

1332. CODEIA was discovered lately by Robiquet in the common muriate of morphia. Ammonia added to a solution of this substance in water precipitates the morphia alone, the codeia being left in the solution, and separated afterwards by crystallization. Codeia has a distinct alkaline reaction; fuses when heated to 300°, does not render nitric acid red; is more soluble in water than morphia.

1333. NARCEIA is another alkali which has been discovered by Pelletier in the watery infusion of opium. White, crystalline, melts about 200°. Its salts are blue when dissolved in a particular quantity of water, the colour changing to violet and red as it is increased.

1334. THEBAIA is the name of another alkaline principle found in opium, which is considered the same as the PARAMORPHINE of Pelletier. White, crystallizes in rhomboidal prisms; soluble in ether; fuses at 266°.

1335. MECONIA, or MECONINE, is another constituent of opium, found in very minute quantity associated with narceia as it is at first procured, being separated from it by ether, and removed by water from the extract made by evaporating the ethereal solution. It is solid, easily fused; resembles fat in appearance, and is soluble in water, alcohol, and ether.

1336. From 40 pounds of opium, Coverba obtained 50 oz. of Morphia.

...	...	$1\frac{1}{2}$	of Codeia.
...	...	1	of Thebaia
...	...	1	of Meconia.
...	...	$\frac{2}{3}$	of Narceia.

1337. Colour produced by agitating the preceding substances with sulphuric acid mixed with a little nitric acid,

Morphia gives a brownish colour.		
Codeia	...	green colour.
Thebaia	...	yellow rose colour.
Meconia	...	turmeric yellow, and then a red colour.
Narceia	...	chocolate colour.

1338. Meconic Acid.—Solid, crystallizable, soluble in water and alcohol, particularly distinguished by the red tint which it gives to iron, similar to that produced by the sulphocyanic acid. By boiling with water it is converted into metameconic acid, a compound lately discovered, and water.

1339. Preparation.—Acetate of lead is added to the watery solution of opium, the meconate of morphia which it contains gives a precipitate of meconate of lead, the acetate of morphia being left dissolved. On washing the precipitate, and mixing with water, it is decomposed by sulphureted hydrogen; the products are sulphuret of lead, water, and meconic acid in solution. From this the meconic acid is procured by evaporation. It may be purified by repeated solution and crystallization.

1340. Antidotes to Opium. Carbonate of potassa has been recommended in cases of suspected poisoning by opium, precipitating carbonate of morphia, which is less active than the meconate of morphia.

1341. To detect opium, when it has been administered as a poison, Dr Christison recommends the contents of the stomach to be acidulated with acetic acid, evaporated to a syrup, and digested in strong alcohol. The alcoholic solution is then evaporated, and the residue dissolved in water; the meconic acid and morphia may be then detected by the test already described in pars. 1327. and 1338.

CHAP. II. QUINA AND CINCHONIA.

1342. Quina.—Eq. 162 = carbon 120 + oxygen 16 + nitrogen 14 + hydrogen 12.—Solid, uncrystallizable, fusible, insoluble in water, soluble in alcohol.

1343. Sulphate of Quina, called technically Disulphate, containing 2 eqs. of quina, and 1 of sulphuric acid; crystallizable, contains 2 or more eqs. of water, according to the temperature to which it may have been subjected after crystallization. Sparingly soluble in cold water, dissolved abundantly by boiling water: phosphorescent when heated.

Alkalis precipitate quina, combining with the sulphuric acid.

Alkaline carbonates precipitate carbonate of quina.

Oxalate, tartrate, and gallate of quina are precipitated, when oxalates, tartrates, and gallates are added to a solution of the sulphate of quina.

1344. Preparation.—Yellow Peruvian bark (Cinchona cordifolia) is boiled in water acidulated with sulphuric acid. The solution of supersulphate of quina thus formed is decomposed by lime. Sulphate of lime is formed, and the quina mixes mechanically with it, as it is precipitated. Alcohol dissolves the quina and leaves the sulphate of lime. The alcohol being evaporated, the quina is procured by itself; neutralized by diluted sulphuric acid, and boiled with animal charcoal to destroy colouring matter, a solution is procured, which gives crystals of the sulphate on evaporation.

1345. Sulphate of quina is often adulterated. EARTHY SUBSTANCES are detected by heating the sulphate in contact with air, remaining after it has been dissipated. SUGAR is detected by dissolving in water the sulphate suspected to contain it, precipitating carbonate of quina by carbonate of potassa, sulphate of potassa remaining in solution along with the sugar, which may be recognised on tasting the solution after the quina has been separated; or the solution may be evaporated to dryness, and the sugar removed by alcohol from the sulphate of potassa. Fatty matters are detected by the action of boiling water, which removes the sulphate, but leaves them undissolved.

1346. Peruvian bark of good quality gives an infusion which precipitates gallic acid (the infusion of galls may be used for this purpose); when there is no vegetable alkali present to combine with the gallic acid, there is no precipitate.

1347. CINCHONIA.—Eq. 154 = carbon 120 + nitrogen 14 + hydrogen 12 + oxygen 8.—Similar in its properties to quina; but differs from it in its equivalent weight and composition, and in crystallizing from its alcoholic solution. It is prepared from the pale Peruvian bark (Cinchona condaminea), by digesting it in dilute muriatic acid, precipitating the cinchonia by an alkali or earth, and dissolving it in alcohol, from which it may be obtained in 4-sided prisms.

1348. KINIC ACID is the substance with which quina and cinchonia are associated in Peruvian bark. A watery infusion of the bark evaporated to a syrup, and treated with alcohol, leaves a viscid mass containing kinate of lime and gum. By solution and crystallization, kinate of lime may be procured in crystals, from which the lime may be separated, by dissolving the kinate in water, and adding sulphuric or oxalic acid. The compounds it forms with the alkalis are very soluble in water.

CHAP. III. STRYCHNIA, &c.

1349. STRYCHNIA.—Eq. 234 = carbon 180 + oxygen 24 + hydrogen 16 + nitrogen 14. Intensely bitter, nearly insoluble in water, very soluble in boiling alcohol, crystallizes in 4-sided prisms. Exceedingly active in its effects upon the animal economy; half a grain kills small animals, as rabbits, in a few minutes. PREPARATION.—*Nux Vomica* is macerated in water, and the solution evaporated to the consistence of an extract; alcohol is added, which dissolves igasurate of strychnia; an alcoholic extract having been made, water is added to dissolve the igasurate and separate fatty matter. By lime-water, the strychnia is precipitated in combination with a portion of brucia. By macerating the precipitate with diluted alcohol, the brucia is dissolved and separated; boiling alcohol dissolves the strychnia, and deposits crystals as it cools.

1350. IGASURIC ACID is prepared by precipitating it from the watery solution of Nux vomica by acetate of lead, decomposing the igasurate of lead diffused through water by hydrosulphuric acid, when insoluble sulphuret of lead is produced, while the igasuric acid is left in solution.

BRUCIA resembles strychnia much, and may be procured from the nux vomica in small quantity, and also from the Brucea antidysenterica. Like morphia, it strikes a deep red tint with nitric acid; and strychnia, which produces this effect, is considered to contain a small portion of brucia.

VERATRIA is the alkaline principle of Colchicum autumnale (Meadow-Saffron), and Veratrum album (White Hellebore).

SOLANIA is procured from the Solanum dulcamara (Woody Nightshade).

DELPHINIA exists in the Delphinium staphysagria (Stavesacre).

EMETINA is obtained from Ipecacuanha.

Other vegetable alkalis have been described, the existence of some of which has not been admitted by all chemists; the principal are the following :—

Names.		*Sources.*
ATROPIA	from the	ATROPA BELLADONNA.
Picrotoxia	......	Cocculus Indicus.
Daturia	......	Datura Stramonium.
Daphina	......	Daphne mezereum.
Digitalia	......	Digitalis.
Conia	......	Conium maculatum.
Hyoscyama	......	Hyoscyamus niger.

ORDER III. OLEAGINOUS AND OTHER COMPOUNDS ABOUNDING IN HYDROGEN AND CARBON.

1351. Some of the compounds in this order contain no oxygen, according to the most recent analysis, as oil of turpentine. Caoutchouc contains scarcely 1 per cent. In general the proportion of oxygen varies from 8 to 14 per cent., the hydrogen from 5 to 15, and the carbon from 70 to 90. The large proportion of their hydrogen and carbon accounts for their great inflammability. Some of them burn with a clear and steady light, producing little smoke, but most of them burn with much smoke, when particular precautions are not taken to insure the combustion of all the carbon. The unconsumed carbon collected for sale is called LAMP BLACK; usually got by the combustion of resinous bodies or tar.

1352. OILS are divided into two classes, Fixed and Volatile Oils.

1353. FIXED OILS are usually obtained by expression. They are not converted into vapour without decomposition, and a temperature above 600° is required for this purpose. Fluid in general at natural temperatures; some are solid, as palm oil. At a red heat they are resolved into inflammable gases and charcoal. Absorb oxygen on long exposure to the air, becoming rancid from the decomposition of mucilaginous matter, and sometimes assuming a waxy consistence. Some, as linseed oil, absorb oxygen rapidly when heated, and acquire the property of drying; in this condition they are used in the preparation of printers' ink and oil paint.

During the spontaneous absorption of oxygen, more especially when the oil is diffused over hemp or cotton, so much heat is produced that combustion ensues; numerous fires have arisen in this manner. Sulphuric and nitric acids decompose oils. Water and alcohol have little effect in general upon them, but *castor oil* and alcohol unite in all proportions. By animal charcoal they may be decolorized. By boiling with water, the offensive odour and taste which they sometimes acquire may be occasionally removed. With sugar or gum they may be suspended in water, forming an emulsion.

1354. These oils contain two proximate principles, discovered by Chevreul, ELAINE and STEARINE. Elaine, called also oleine, is much more fluid than stearine, remaining in this condition when the stearine is separated in the solid form by a reduction of temperature to 25°.

1355. Many varieties of stearine have been discovered, one of which is termed MARGARINE, predominating in fixed vegetable oils, while the STEARINE is observed chiefly in animal oils and fats.

1356. Fixed oils are converted into a peculiar kind of acids and GLYCERINE, when heated with the fixed alkalis and water. The acids uniting with the alkalis constitute SOAP; the glycerine remains in solution. It is there associated with a portion of free alkali, which must be neutralized by sulphuric acid; on evaporating the whole to dryness, alcohol may be used to extract the glycerine in solution, affording a syrupy fluid when concentrated. Glycerine is soluble in water and alcohol, has a sweet taste, does not crystallize, nor does it enter into the vinous fementation.

1357. MARGARIC, STEARIC, and OLEIC acids, are the names applied to these acids, being formed from the margarine, stearine, and elaine of the oil. These acids may be procured by decomposing the different kinds of soap in which they abound (which may be termed MARGARATES, STEARATES, and OLEATES of the alkalis with which the acids may be combined) by sulphuric acid, an alkaline sulphate being left in solution, while the insoluble acid is separated.

1358. MARGARON, STEARON, and OLEON, are the names given by Bussy to the products obtained by distilling a mixture of margaric, stearic, and oleic acids with 1-4th of their weight of lime, and purifying the fatty or oily compound obtained by repeated solution in alcohol. Margaron and stearon can be crystallized, but oleon is liquid at natural temperatures.

1359. SOAP is decomposed by mineral and many vegetable acids, which unite with the alkali, separating the oleaginous acid. Alkaline earths form a white precipitate with the acid, leaving the alkali in solution. Earthy salts produce a double decomposition, the oleaginous acid and earth being precipitated, while an alkaline salt remains in solution; thus margarate of potassa and sulphate of lime precipitate margarate of lime, sulphate of potassa remaining in solution. A solution of soap in alcohol is much employed as a test of the presence of earthy salts in water. With water absolutely pure the transparency is not impaired; the larger the quantity of earthy salts in solution, the greater the precipitation.

1360. The soaps prepared with potassa are generally soft, but those made with soda are usually hard. Silicated potassa is mixed with common soap, constituting the more common variety of flint soap. See 918.

1361. VOLATILE OR ESSENTIAL OILS rise in vapour at 212° when mixed with water, and are usually prepared by distillation; they are volatilized without water at a temperature about 300°. Odour pungent, and generally fragrant; taste acrid; soluble in alcohol, very sparingly soluble in water. By long exposure to oxygen they are converted into a resinous matter. They dissolve sulphur and phosphorus when these are boiled along with them. Sulphuric and nitric acids act violently on them; by nitric acid they are often inflamed. Alkalis and earths have little action with the volatile oils. Fixed oils combine with them, and are often added to adulterate them; they are detected by heating a little on paper, a greasy stain being left where a fixed oil is present. Alcohol is detected by water, which uniting with it leaves a smaller volume of oil.

1362. RESINOUS SUBSTANCES are procured from the resinous juice that exudes from different kinds of plants. They are highly inflammable, decomposed in general with the same phenomena as the oils, insoluble in water, soluble in alcohol, ether, volatile oils, and solutions of the fixed alkalis; with the latter they form soap of different kinds.

Common resin is procured by exposing common turpentine, the resinous juice of the fir tree, to heat with water; oil of turpentine is separated by distillation, and the resin left.

1363. CAMPHOGEN OF CAMPHENE, called also ESSENCE OF TURPENTINE, is procured by distillation from common turpentine, and may be regarded as oil of turpentine in its highest state of purity. Its symbol has been very differently stated; H^8C^{10} has been considered to represent it correctly. Transparent, colourless, inflammable, burning with a dense smoke. It boils at 312°. Alcohol can dissolve a small quantity; with water it forms a crystalline compound termed HYDRATE OF CAMPHENE. With 1 eq. of oxygen it forms camphor, and with a larger quantity camphoric acid is produced.

1364. Common oil of turpentine is seldom obtained free from oxygen, which it absorbs from the air.

1365. CAMPHOR has been usually regarded as a concrete volatile oil, but may now be more correctly termed OXIDE OF CAMPHENE. Solid, white, crystalline, scarcely soluble in water, but suspended in it by means of magnesia it forms an emulsion. Soluble in alcohol, acetic and nitric acids; precipitated from these solutions by water: very inflammable: melts at 288°, boils at 400°. CAMPHORIC ACID is formed by treating it with nitric acid.

1366. ARTIFICIAL CAMPHOR is a crystalline compound of camphene and muriatic acid, formed most easily by the action of the acid on camphene at a low temperature.

1367. BENZULE is the name applied to a peculiar principle which exists in the oil obtained by distillation with water from the fixed oil of the bitter almond prepared by expression. It has not hitherto been procured in a free state, but may be separated from one substance and transferred to another in numerous combinations. Symb. $O^2C^{14}H^5$, or Bz, which may be adopted as a provisional symbol. The fixed oil of the bitter almond is decomposed in the preceding process; the volatile oil which is collected in the receiver is a HYDRURET OF BENZULE. It is separated from hydrocyanic acid by distilling it from a solution of potassa mixed with muriate of iron, the precipitated oxide of iron retaining the acid. By distillation from quicklime in powder, it is obtained free from water.

1368. When oxidated, benzule forms BENZOIC ACID. Chlorine, iodine, bromine, sulphur, and cyanogen, form compounds with it. BENZAMIDE is a name given to another new principle formed during the action of benzule with chlorine and ammonia; it may be considered a compound of hydruret of benzule and nitrogen. BENZOINE is the crystalline compound that is deposited from oil of bitter almonds and other oils, when kept in contact with potassa for some weeks.

1369. COUMARINE is obtained ftom the Tonqua bean, and is analogous to the essential oils; its properties have not been minutely investigated.

1370. CAOUTCHOUC, the concreted juice of the Ficus elastica, and some other trees. Many of its properties are familiarly known, as it constitutes the common Indian rubber. Insoluble in water and alcohol; dissolved by turpentine, naphtha, ether, and in the fluid obtained by heating the caoutchouc itself. Made into the consistence of paste by naphtha or other solvents, it is used in the preparation of WATERPROOF CLOTH. The fluid prepared by distillation from caoutchouc is particularly distinguished by its great volatility, and low specific gravity, not exceeding 0.64 when highly rectified.

1371. Several specimens of it, for which I have to thank Mr Enderby of London, who manufactures large quantities of it, have enabled me to prepare a blowpipe lamp of great power. The flame is produced by igniting a stream of air passed through the volatile liquid, which produces an inflammable vapour from the large quantity of it which is dissolved.

1372. WAX.—Solid, brittle, white when pure; melts at 150°; insoluble in water; dissolved in small quantity by ether. Alcohol dissolves MYRICIN, leaving CERIN, the other principle of which wax is composed. Soluble in fixed and volatile oils. Little affected in general by acids; forms soaps when heated with solutions of the fixed alkalis.

1373. GUM RESINS.—This term is applied to a number of concrete vegetable juices, as gamboge, guaiac, aloes, and assafœtida, which contain various proportions of resin, gum, and other vegetable principles.

1374. BALSAMS are the viscid liquids or solids produced by different vegetable juices containing benzoic acid.

1375. AMBER is regarded by Berzelius as a CONCRETED BALSAM; it contains an oily matter, succinic acid, and other proximate principles of less importance.

1376. COAL, and the other bituminous substances mentioned in the succeeding paragraphs, are all regarded as of vegetable origin.

1377. COMMON COAL contains carbon, oxygen, hydrogen, and nitrogen, with variable proportions of the alkaline, earthy, and common metallic salts; sulphuret of iron is frequently mixed with it. The following are the more familiar varieties:—

1378. I. SPLINT COAL, which derives its name from the splintery fracture it presents.

1379. II. CAKING COAL, which undergoes a partial fusion, and forms a cake when heated.

1380. III. CHERRY COAL has a slaty fracture, and is more easily broken than splint coal.

1381. IV. CANNEL COAL, or PARROT COAL, gives much gas when heated, and was formerly used as a substitute for candles.

1382. V. GLANCE COAL, or ANTHRACITE, contains no bituminous matter, consisting almost entirely of carbon. It gives no gas when heated, but burns with a fixed light.

1383. VI. BROWN COAL, BITUMINOUS WOOD, or SURTURRAND, has a brownish colour, and resembles peat in the phenomena produced during its combustion.

1384. VII. JET, which is so much employed in making black ornaments, is another variety of coal.

1385. NAPHTHA, as it occurs in native springs, is generally purer than the naphtha prepared from coal tar, and less liable to be contaminated with oxygen. Some varieties are considered to be very similar to camphene, the pure essence of turpentine, and others to differ from it. It is generally associated in various proportions with other bituminous matters, and is seldom procured transparent and colourless. NAPHTHA, PETROLEUM, MINERAL TAR, MINERAL PITCH, and MINERAL CAOUTCHOUC, may all be regarded as varieties of bituminous matter formed from coal. ASPHALTUM is a solid bituminous compound, much employed in making varnishes. RETINASPHALTUM is the name given to a peculiar bituminous matter, associated with some varieties of brown coal.

Compounds procured from the products obtained by the destructive distillation of Wood and other substances.

1386. TAR is the impure resinous juice extracted from the fir tree by heating it artificially; it is black from the charcoal mechanically mingled with it. Subjected to heat without access of air, acetic acid, water, and many inflammable products, are expelled, the resinous matter alone being left blackened by the charcoal. Coal subjected to heat, besides the gaseous products and water charged with ammonia, gives a tarry matter, resembling in many of its properties that which is procured from wood.

1387. PYROXILIC SPIRIT is the name given to the volatile liquid which is formed in considerable quantity along with acetic acid, when wood is subjected to heat, and the air carefully excluded. It has unfortunately received a variety of names, as PYROLIGNEOUS ETHER, HYDRATE OF CARBOHYDROGEN, HYDRATE OF METHYLENE. Its symbol is OCH^2, or ·H+HC; the term HYDRATED HYDRURET OF CARBON would express its composition in unison with the nomenclature adopted in this work. Eq. 16.

1388. Liquid; sp. gr. .798. Boils at 151°. Burns with a pale lambent flame, resembling that of alcohol. Its odour is strong and peculiar. It is used principally as a substitute for alcohol in spirit lamps.

1389. A number of new compounds have lately been formed by the action of different acids subjected to distillation along with pyroxilic spirit, in which 1 eq. of carbohydrogen is united with 1 eq. of acid, as in the sulphate, nitrate, hydrochlorate, hydriodate, &c. of carbohydrogen. When 1 part of pyroxilic spirit is distilled with 4 of aqueous sulphuric acid, a new gas is procured, isomeric with alcohol; sulphurous and carbonic acids, which are mixed with it at first, are removed by potassa. This compound is termed DIHYDRATE OF CARBOHYDROGEN. Symb. ·H+2HC.

1390. PYROACETIC SPIRIT.—Symb. OC^3H^3, or O+3HC.—Eq. 29.—It may be obtained most easily by heating a number of acetates, the acetic acid being resolved entirely in some cases into pyroacetic spirit and carbonic acid. It is a transparent, colourless, volatile, and inflammable liquid. It enters into combination in all proportions with water, alcohol, ether, and some volatile oils.

1391. PARAFFINE, or PETROLINE, discovered by Dr Reichenbach, and by Dr Christison, is procured by distillation of the tar from beech-wood, and from the petroleum of Rangoon, impurities mixed with it being decomposed by heated sulphuric acid; the paraffine is separated in a solid form. White, solid, tasteless, inodorous, soluble in alcohol, turpentine, and naphtha; crystallizable; fuses at 111°. Particularly distinguished by resisting the action of acids, alkalis, and other energetic chemical agents. It burns with great brilliancy, and without smoke. Its name indicates its feeble tendency to enter into combination, being derived from the words *parum* and *affinis*.

1392. EUPION, from the Greek ευ, well, and πιων, fatty or greasy. Liquid, boils at 339°, soluble in alcohol, ether, and oil of turpentine. Obtained by distillation from the tar of animal matters; purified by distillation with sulphuric acid; paraffine is separated in the solid form by reduction of temperature. This principle, and the four succeeding compounds, were also discovered by Dr Reichenbach.

1393. KREOSOTE, from κρεας, flesh, σωζω, I save. Composed of carbon, oxygen, and hydrogen. Liquid, transparent, colourless, inflammable, burning with a smoky flame. Sp. gr. 1.037. Boils at 397°. Odour very penetrating and resembling that of smoked meat; very antiseptic, being the principle on which the antiseptic powers of smoke and pyroligneous acid depend. Soluble in alcohol, ether, naphtha, and acetic acid. Decomposed by nitric and sulphuric acids; combines with alkalis and earths; coagulates albumen, from which property its antiseptic virtues are probably derived. It has been employed much in toothache, being applied in minute quantity where the tooth is decayed. It has been used in solution for the purpose of curing meat.

1394. Prepared by repeated distillation from the oil distilled from wood tar. Carbonate of potassa is added to neutralize acetic acid; the acetate having been removed, the distillation is commenced. Phosphoric acid is then employed to separate ammonia, and potassa is subsequently combined with it, that the eupion associated with it may float upon the surface and be removed; the potassa is separated afterwards by sulphuric acid, and the kreosote again distilled when the sulphate of potassa has been separated.

1395. PITTACAL is the name given to one of the principles of tar, which gives a blue colour to baryta in solution; this term is derived from πιττα, pitch, and καλοσ, beautiful.

1396. PICAMAR is the bitter principle of tar. Picamar is derived from pix and amarus. Obtained from distilled oil of tar, having a great specific gravity.

1397. CAPNOMOR, from καπνοσ, smoke, and μοιρα, a part. Procured from the heavy oil of tar by heating it with potassa, which decomposes the picamar, the pittacal, and the creosote, the capnomor floating on the surface. On the addition of sulphuric acid, any eupion is separated, the capnomor combining with the acid. Carbonate of potassa may be added to neutralize the sulphuric acid, after which the capnomor is to be purified by distillation. This liquid is transparent and colourless; its odour is fragrant. Some specimens prepared in the manner described turn the vegetable blues to a green.

1398. BENZINE is the name applied by Mitscherlich to the bicarburet of hydrogen, which is procured by heating benzoic acid with lime. His nitro-benzide and sulpho-benzide are compounds formed by the action of nitric and sulphuric acids on benzine.

1399. BENZONE is a volatile fluid which was procured by Peligot by heating dry benzoate of lime. Its symbol is $OC^{13}H^{5}$.

ORDER IV. LIGNIN, SUGAR, STARCH, GUM-ARABIC, GLUTEN, AND TANNIN.

1400. The four first principles included in this order are considered to contain oxygen and hydrogen in the proportions necessary to constitute water. All of them occur abundantly in the vegetable kingdom.

1401. LIGNIN, or WOODY FIBRE.—The most abundant of vegetable principles, constituting upwards of 90 per cent. of the different kinds of wood. Obtained pure after saw-dust has been acted on by alcohol, water, and diluted acids. Insoluble in water. Nitric acid produces with it oxalic and other acids. By strong sulphuric acid it is decomposed, carbon being deposited, while sulphurous acid and water are produced; with weak sulphuric acid a saccharine matter may be formed.

1402. SUGAR.—Symb. OCH = oxygen 8 + carbon 6 + hydrogen 1.—Eq. 15? Several consider that 45 (15 × 3) expresses more correctly the equivalent of sugar. Solid, white, inodorous, decomposed by heat, producing much flame when exposed to the air; does not deliquesce when pure; soluble in water and alcohol; decomposed by sulphuric acid, charcoal being deposited; nitric acid produces oxalic and other acids. Combines with alkalis, earths, and some metallic oxides.

1403. Prepared from the juice of the sugar-cane by evaporation and crystallization; lime water is added to neutralize vegetable acids mixed with it. Albuminous fluids, such as blood, are added to entangle and remove impurities; charcoal to destroy colouring matter; and, according to the most approved plan, the syrup is concentrated for granulating (crystallizing) by evaporation in vacuo.

1404. MOLASSES are composed principally of sugar, with various vegetable impurities, which prevent it from crystallizing.

1405. HONEY consists of sugar, with a portion of uncrystallizable saccharine matter probably analogous to molasses.

1406. LIQUORICE and MANNA; each contains a peculiar kind of saccharine matter called GLYCYRRHIZINE and MANNITE, associated with other vegetable principles.

1407. Starch.—White, tasteless, inodorous; insoluble in cold water, alcohol, and ether; dissolved by boiling water, gelatinizing when a small quantity is added; particularly distinguished by the deep blue coloured compound it forms with iodine in cold water,—the hydrated iodide of starch; when quite free from water, the iodide is yellow; in hot water no coloured compound appears; the blue coloured iodide disappears in boiling water, but, if cooled quickly, the colour is restored. If the blue coloured compound be boiled for a short time, hydriodic acid is formed, and the blue colour is not restored on cooling, unless chlorine or some other substance be added to liberate the iodine. When starch has been gelatinized in water, it is converted into amidine, which is soluble in cold water. The same effect may be produced by heating dry starch; subjected to a higher temperature, so as to acquire a reddish-brown tint, it becomes more soluble in cold water, forming the compound termed British gum. Starch does not precipitate silicated potassa, nor produce mucic acid with nitric acid.

1408. Starch is converted into sugar during the germination of seeds, as may be well seen in the operation of malting barley: it is also converted into sugar by the action of dilute sulphuric acid.

1409. Starch or fecula is procured mechanically from the potato and other roots, from wheaten flour, and other vegetable products. Potato Flour, Indian arrow-root, British and Foreign tapioca, sago, and cassava, are composed almost entirely of starch.

1410. Gum is a term which now includes several different principles.

1411. Gum-Arabic contains a minute portion of nitrogen, and a little lime, in addition to the usual elements of vegetable matter. Solid, soluble in water; precipitated from its aqueous solution by alcohol; insoluble in alcohol and ether; decomposed by sulphuric acid; produces mucic acid with nitric acid; precipitated by subacetate of lead and silicated potassa. With water a mucilaginous fluid is formed. A strong solution of borax immediately renders 4 times its bulk of a strong solution of gum very viscid; sugar added to the viscid mass immediately renders it liquid.

1412. Cherry-tree Gum; analogous to gum-arabic in composition, but insoluble in cold water. Called cerasin by Guerin; acquires the same properties as gum-arabic when heated. The mucilage of linseed resembles the solution of gum-arabic in water, but differs from it in some less important properties.

1413. Gum-Senegal is considered similar to gum-arabic.

1414. Gum-Tragacanth is composed of a compound similar to gum-arabic, with a large quantity of bassorine, and a portion of starch.

1415. Bassorine, from gum-bassora, gum-tragacanth, or salep. It is converted by long continued boiling into a substance similar to gum-arabic.

1416. Pectin is the name given to the gelatinous matter which abounds in many kinds of fruit, as in the gooseberry. It forms a mucilage with water, not so adhesive as gum, and is insoluble in alcohol. Alkalis and alkaline earths convert it into pectic acid, with which they combine.

1417. Gluten is the most nutritious part of wheaten flour, and exists also in most other kinds of grain in smaller quantity; it resembles animal more than vegetable substances, containing nitrogen, besides carbon, oxygen, and hydrogen. Gluten is a stiff and tenacious solid, possessing ductility and elasticity, so that it can be spread into a thin membranous sheet. It is procured by making wheaten flour into a paste, and removing the starch, saccharine and other matters, by washing it with water. It may be resolved into two principles, gliadine and zimome.

1418. Gliadine is separated in solution from gluten by alcohol. Zimome, the other component part of gluten, is insoluble in alcohol, and is distinguished by the bluish-green colour it gives when triturated with powdered gum-guaiac, and exposed to the air.

1419. Gluten is found in small quantity, or not at all, in inferior kinds of flour. Gum-guaiac has been proposed as a test for ascertaining the quantity of gluten in different kinds of flour.

1420. Vegetable Albumen is coagulable by heat, and bears a great resemblance to animal albumen. It has been found in the sweet and bitter almond, and in a number of different kinds of seed.

1421. Tannin exists in a great variety of vegetable products, but is found most abundantly in some vegetable extracts, as catechu and kino, in gall-nuts (the excrescences that grow on the oak), and in the bark of the oak.

1422. It is colourless and inodorous. Taste astringent. Soluble in water, alcohol, and ether; gives a deep bluish-black precipitate with persalts of iron, but has little action with the protosalts. It approaches, therefore, in its characters to gallic acid, but is distinguished from it by precipitating gelatine. When highly purified, it is said to redden litmus, and decompose carbonates with effervescence, so that it is now proposed to place it along with the vegetable acids, under the name of tannic acid. Exposed to the air in solution, oxygen is absorbed, carbonic acid evolved, and it is converted ultimately into gallic acid.

1423. Tannin is precipitated from the infusion of galls by protomuriate of tin; combining with the oxide, and forming an insoluble compound. When the precipitate is diffused through water, and decomposed by sulphureted hydrogen, sulphuret of tin is formed, and tannin remains in solution.

1424. Artificial Tannin is produced by the action of nitric acid on charcoal, or on coal, resin, indigo, and other substances containing much charcoal. It resembles tannin much, but is not decomposed like it by nitric acid.

ORDER V. COLOURING MATTER.

1425. A very great number of proximate principles may be referred to this order. The following are those most interesting to the student of chemistry.

1426. *Colouring matters used for testing the presence of acids and alkalis.*

I. By infusing in water the common red cabbage, as it is familiarly termed, a solution of the colouring matter is procured, which is blue; it is turned to a bright red by acids, and becomes of a brilliant green by the action of alkalis. The colouring matter of violets may be used for the same purpose.

II. Litmus is the blue colouring matter prepared from the Lichen Rocella. It is more easily affected by acids than the colouring matter of cabbage, but is not turned to a green by alkalis. If previously reddened by acids, it may be used for detecting alkalis, the original blue tint of the litmus being restored.

III. Turmeric is of a yellow colour, and detects alkaline substances by producing with them a brown coloured compound. It is not affected by acids. If previously rendered brown by an alkali, it may be employed for detecting acids, the alkali being then neutralized, and the yellow colour of the turmeric restored. Some neutral salts affect turmeric in the same manner as alkalis.

1427. Many colouring matters have a great attraction for metallic oxides, combining with them when they are separated from solutions in which both the colouring matter and the oxides have been previously dissolved; thus, when potassa is added to a solution containing bipermuriate of tin and litmus, the potassa unites with the muriatic acid, litmus and peroxide of tin falling down in combination. The precipitates procured in this manner are usually called lakes.

1428. SUBSTANTIVE COLOURING MATTERS include those which have so strong an attraction for cloth, that they can attach themselves to it permanently, and without the action of any other substance, as indigo.

1429. ADJECTIVE COLOURING MATTERS are those which have a feeble attraction for cloth, and are fixed on it only by the medium of some other matter, which can combine both with the colouring matter and the cloth; this is termed a MORDANT or BASIS. The usual mordants are alumina and the oxides of tin and iron.

1430. Colouring matters are BLEACHED most powerfully by CHLORINE, which decomposes them by combining with their hydrogen, or by taking hydrogen from water, and liberating oxygen, which, in a nascent state, may be equally effectual in decomposing them. In some of the decolorizing compounds of chlorine, a new compound, HYPOCHLOROUS ACID, is considered to be in union with a base; and when they act upon colouring matter, they have been regarded as operating more especially if not entirely by the oxygen of the hypochlorous acid and of the oxide, a metallic chloride being left. The CHLORIDE OF LIME is the bleaching powder in common use; 2 eqs. of water are usually combined with it. SULPHUROUS ACID, and some other acids diluted with water, are much employed in decolorizing different substances.

1431. For RED DYES, Brazil wood, bar-wood, lac, archil, madder, and cochineal, are the principal colouring matters in common use. The cochineal is procured from an insect, which is believed to derive its colouring matter from a particular vegetable principle upon which it feeds. Oxide of manganese and catechu are much used for browns.

1432. YELLOW DYES are procured principally from saffron, hiccory, quercitron bark, turmeric, fustic, and annatto. A buff is obtained with peroxide of iron.

1433. BLACK DYES are made with the same materials as writing ink, as gallic acid, tannin, and peroxide of iron; logwood and madder also give a black along with peroxide of iron.

1434. BLUE DYES are commonly prepared with Prussian blue or INDIGO. Three different colouring principles have been detected in indigo, INDIGO BLUE, INDIGO RED, and INDIGO BROWN. The first of these is the most important, and is left in a pure form when indigo has been digested successively with diluted sulphuric acid, a solution of potassa, and alcohol.

1435. INDIGO BLUE sublimes at 550°, producing a reddish violet-coloured vapour, which condenses in crystals. Insoluble in water and alkaline solutions; boiling alcohol dissolves a minute quantity. Strong sulphuric acid is its proper solvent. When the solution is decomposed by diluting it with water, a portion of the indigo is deposited in combination with water, forming a hydrated compound, called PHENECIN, or INDIGO PURPLE. Berzelius has also described two new compound acids, INDIGO SULPHURIC ACID, and INDIGO HYPOSULPHURIC ACID, which are formed by the action of sulphuric acid on indigo blue. The INDIGOTIC ACID already mentioned, is procured by boiling indigo in dilute nitric acid. The carbazotic acid is prepared by the action of indigo with stronger nitric acid.

Indigo is deoxidated by a number of substances; it can then unite with lime or potassa, and the compound forms a yellow solution with water. Sulphate of iron mixed with lime precipitates oxide of iron, which produces this effect. Sulphuret of arsenic dissolved in a solution of potassa, acts in the same way. INDIGOGEN is the name applied to deoxidated indigo. It attracts oxygen quickly on exposure to the air, and indigo blue is again produced: at first a green appears, from the mixture of the blue and yellow.

1436. Indigo blue dissolved in sulphuric acid, is employed for ascertaining the quantity of chlorine in chloride of lime; the larger the quantity of chlorine in the chloride the greater the quantity of indigo blue which it can decolorize.

1437. By combining red, yellow, blue, or black colouring matters with each other, mechanically or chemically, all other tints may be produced; and by varying the strength of the colouring matter, or the strength of the mordant, different shades of the same colour may be had.

CHROMULE is the term applied to the various coloured principles obtained from the leaves and flowers of plants.

CHLOROPHYLE is the name given to the green colouring matter of vegetables; in the autumn it is reddened by the production of acid.

1438. In a paper read lately before the Royal Society by Dr Hope, he adduced numerous experiments to prove that there are two varieties of vegetable colourable matter, or CHROMOGEN, as he has designated them, which are acted upon by acids and alkalis in producing red, yellow, or green tints. The one he has termed ERYTHROGEN, as it produces a red colour with acids; the other he has denominated XANTHOGEN, from its affording a yellow compound with alkalis. These two principles occur together in red and blue flowers, and in the leaves of a few plants, presenting the former of these tints. All green leaves, all white and yellow flowers, and white fruits, contain xanthogen alone. Litmus abounds in erythrogen, but has no xanthogen. The chromates having different tints, are regarded as peculiar vegetable principles, sometimes intimately blended, or chemically united with chromogen; they are also occasionally compounds of chromogen with acids or alkalis.

ORDER VI. MISCELLANEOUS VEGETABLE PROXIMATE PRINCIPLES.

1439. The different principles contained in this order being comparatively less interesting than those already described, it will be sufficient in this place to mention their names, and the sources from which they are derived.

AMYGDALINE, from the bitter almond.

ASPARAGINE, from asparagus.

BRYANINE, from bryony root.

CAFFEINE, from coffee.

CATHARTINE, from senna.

COLOCYNTINE, from the colocynth.

COLUMBINE, from columbo root.

ELATINE, from elaterium.

FUNGINE, from mushrooms.

GENTIANINE, from gentian.

HORDEIN is found in large quantity in barley, and is converted into starch by the saccharine fermentation, as during the process of malting.

INULINE, from Inula Helenium, or Elecampane.

LUPULINE, from the hop.

MEDULLINE forms the pith of the sunflower.

OLIVINE, from olive oil.

PIPERINE, from pepper.

PLUMBAGINE, from the Plumbago Europæa.

PHLORIDZINE is deposited in crystals from the solution procured by boiling the bark of the apple.

POLLENINE, from the pollen of tulips.
RHEINE ?, from rhubarb.
RHUBARBARINE ?, from rhubarb.
SALICINE, from the willow.
SAPONINE, from Saponaria officinalis.
SARCOCOLL, from the Penœa sarcocolla.
SCILLITINE, from Scilla maritima.
SENEGINE, from Polygala Senega.
SINAPISINE, from mustard.
SUBERINE, from cork.
ULMINE, from the juice of the elm and other trees.

ORDER VII. ALCOHOL AND THE PRODUCTS FORMED FROM IT. SACCHARINE AND VINOUS FERMENTATION.

1440. SACCHARINE FERMENTATION.—When starch is moistened with hot water and exposed to the air, a portion of sugar is produced. This change also takes place in some natural operations, as in the process of germination. The production of sugar during the ripening of fruit is also considered an instance of saccharine fermentation; in this case, the sugar is developed by the decomposition of citric acid and mucilaginous matter, independently of the action of the air.

1441. VINOUS FERMENTATION.—Sugar is the only substance which enters readily into a state of vinous fermentation. Starch is also said to be susceptible of this change, but the more general opinion is that it is converted previously into sugar. The sugar must be dissolved in a certain amount of water, and the temperature should be between 60° and 70°, or between 70° and 80°, if it be required that fermentation proceed rapidly. At a higher temperature, vinegar is apt to be formed instead of a vinous liquor. At a low temperature the fermentation proceeds more slowly, or is altogether arrested.

1442. YEAST is the viscid, flocculent matter that is formed during vinous fermentation, collecting generally as a froth or scum on the fermenting liquid. It is composed of oxygen, hydrogen, carbon, and nitrogen, and may be kept without change when dried; it resembles gluten in composition and in some of its properties. Particularly distinguished by its power of promoting vinous fermentation. A small quantity is added in general during the preparation of vinous liquors.

1443. Gluten, albumen, gelatine, and other vegetable or animal properties, can also act as a yeast in forming fermentation.

1444. All vinous liquors, as ale, porter, cider, perry, and wine, contain essentially a portion of alcohol, produced during fermentation by the decomposition of sugar, independent of the action of the air; 45 parts of sugar contain 24 of oxygen, 18 of carbon, and 3 of hydrogen, and they are resolved into 23 of alcohol and 22 of carbonic acid.

Fig. 64.

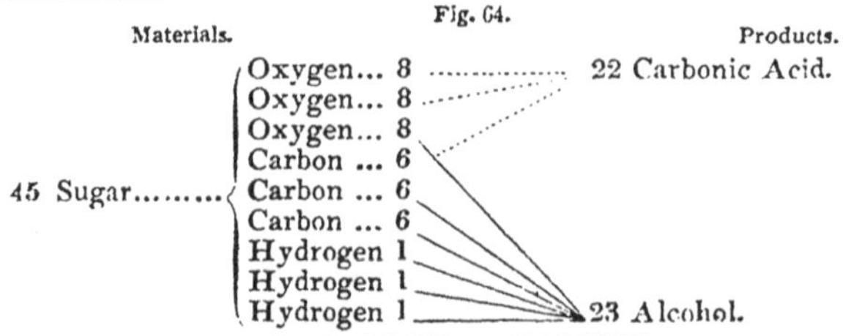

Symb. $3OCH = :C\&OC^2H^3$.

1445. During fermentation, the solution becomes turbid from the production of yeast, the temperature increases, and carbonic acid gas is disengaged. A small portion of yeast is decomposed. From the juice of the grape containing tartaric acid and potassa, these separate in the form of crude tartar (bitartrate of potassa) as the fermentation proceeds.

1446. The nature of the alcoholic compound produced is modified much by the extent to which the fermentation is allowed to proceed, the quantity of mucilaginous, saccharine, and other vegetable principles that may be left in solution, and the proportion of water originally added to the sugar, or associated with it in the juice employed. The stronger wines contain from 18 to 25 per cent. of alcohol; the weaker wines from 12 to 18; and ale, porter, cider, and perry, from 4 to 10.

1447. Alcohol.—Symb. OC^2H^3 = oxygen 8 + carbon 12 + hydrogen 3.—Eq. 23 by W, □ by V.—Sp. gr. 0.796.—Liquid, transparent, colourless. Odour fragrant and penetrating; taste hot and pungent. Boils at 173.5, when its sp. gr. is .810. Has been cooled to —91 without freezing. Inflammable, burns with a pale lambent flame, producing carbonic acid and water; much used for producing a small and steady smokeless flame, which gives considerable heat. Dissolves many vegetable acids, common and vegetable alkalis, resins, camphor, and volatile oils. With water, it combines in all proportions, heat and condensation accompanying the combination. The greatest degree of condensation is observed when 1 eq. of alcohol combines with 3 of water. Proof spirit contains equal weights of alcohol and water. Sp. gr. 0.917. Most acids decompose it, producing a particular class of compounds. The value of alcohol is determined by its sp. gr. the lighter the purer.

Table of the specific gravity of different mixtures of Alcohol and Water, from the Records of Science.

Atoms of		Weight of		Sp. gr. of Mixture.
Alcohol.	Water.	Alcohol.	Water.	
1	0	23	0	.79460
4	1	92	9	.81793
3	1	69	9	.82598
2	1	46	9	.83843
1	1	23	9	.86726
1	2	23	18	.90420
1	3	23	27	.92662
1	4	23	36	.94118
1	5	23	45	.95090
1	6	23	54	.95763
1	7	23	63	.96243
1	8	23	72	.96597
1	9	23	81	.96871
1	10	23	90	.97092

1448. Prepared by distillation from vinous liquids; it is formed in these during fermentation, when 45 parts of sugar give 23 of alcohol and 22 of carbonic acid. The fermented infusion of malt gives whisky by distillation. The fermented solution of molasses produces rum. The fermented juice of the grape affords brandy. Whisky, rum, and brandy, contain from 50 to 54 per cent. of alcohol. By distillation, the alcohol, being more volatile than the water, rises first, carrying along with it, however, a portion of water.

By a second distillation, alcohol is procured with a less proportion of water, forming the common spirit of wine; and by a third distillation it is procured still stronger, constituting RECTIFIED SPIRIT OF WINE. But alcohol is not produced absolutely free from water, unless other processes be resorted to. CARBONATE OF POTASSA can remove all the remaining water, forming a heavy solution, on which the lighter alcohol floats. CHLORIDE OF CALCIUM and other substances retain all the water, when the alcohol is mixed with them; they are separated by distillation. Mr Graham has prepared very strong alcohol by placing a cup containing spirit of wine over fragments of quicklime in the receiver of an air-pump and exhausting it; little alcohol evaporates, as its vapour is not condensed by lime, but all the water evaporates, and its vapour is absorbed by the lime.

1449. ETHERS are the volatile and fragrant fluids produced by the action of acids and alcohol.

1450. SULPHURIC ETHER.—Symb. OC^4H^5.—Eq. 37.—By V □.—Sp. gr. 0.632. Liquid; very volatile, boils at 98°; inflammable. Detonates with great violence when its vapour is mixed with oxygen, if a lighted match be applied. Combines sparingly with water; unites with alcohol in all proportions, dissolves camphor, resins, many fatty matters, volatile oils, caoutchouc, narcotine, and other vegetable principles. Used often to produce great cold by its rapid spontaneous evaporation.

1451 PREPARED by the action of aqueous sulphuric acid and alcohol. Sulphovinic acid is at first formed, and then decomposed; 1 eq. of ether is separated by distillation from 2 eqs. of alcohol; they lose altogether 1 eq. of oxygen and 1 of hydrogen, which combine in the form of water with the acid, while the rest of the elements form the ether. 2 eqs. of alcohol = 1 eq. sulphuric ether and 1 eq. water. Symb. $2OC^2H^3 = OC^4H^5$&·H.

1452. Often contaminated, as prepared at first, with a portion of sulphurous acid and free alcohol. The sulphurous acid is removed by the action of potassa, which retains it on a second distillation. The alcohol is separated by water, which forms with it a heavy fluid, while the lighter and purified ether floats above. A small quantity of ether combines with the water.

1453. SULPHOVINIC ACID has been regarded as a compound of sulphuric acid and 2 eqs. of hydruret of carbon; according to this view its symbol is $\vdots S+2HC$. It has been lately stated to be a compound of alcohol and sulphuric acid. It is procured when oil of wine is heated; or by precipitating baryta from sulphovinate of baryta by sulphuric acid. A solution of sulphovinate of baryta is procured by mixing equal weights of alcohol and sulphuric acid, and adding baryta till the liquid is neutral. Sulphate of baryta is precipitated, and sulphovinate of baryta remains in solution. The compounds of this acid are termed sulphovinates.

1454. OIL OF WINE is composed of 1 eq. of sulphuric acid and 4 of hydruret of carbon, according to some analyses. It is the oily fluid that appears in the neck of the retort on the preparation of sulphuric ether.

1455. NITRIC ETHER (NITROUS ETHER) is produced by the action of equal weights of nitric acid and alcohol, *the acid being added in small successive quantities to the alcohol*, and the mixture cooled after each addition, to prevent the violent action that would otherwise ensue; oxide of nitrogen, binoxide of nitrogen, and carbonic acid, are evolved. It collects on the surface of the mixture, and is cautiously withdrawn. When too freely exposed to the air, it evaporates with rapidity.

1456. Another process consists in mixing aqueous sulphuric acid with alcohol, cooling the mixture, and pouring it upon nitre placed in a retort. The sulphuric acid decomposes the nitre, uniting with the potassa, and liberating nitric acid, which produces nitric ether as it acts upon the alcohol. The nitric ether is separated by distillation.

1457. The precise constitution of this ether is not so generally agreed upon as that of sulphuric ether. It has been regarded as a compound of hyponitrous acid and ether. Symb. $\dot{\dot{\dot{}}}N+OC^4H^5$. It is always acid as prepared at first, but may be obtained neutral by distillation from lime.

1458. SPIRIT OF NITRIC (or NITROUS) ETHER is prepared with a much larger quantity of alcohol, one of acid being mixed with three of alcohol. It is a compound of the undecomposed alcohol and the ether, produced by the action of the nitric acid. From the extent to which the acid is diluted, the mixture may be made at once, taking care to add the acid to the alcohol, and distilling till a quantity of liquid is obtained equal to the alcohol employed.

1459. HYDROCHLORIC or MURIATIC ETHER, is a compound of 1 eq. of hydrochloric acid and 4 of hydruret of carbon. Symb. HCl+4HC. Liquid, colourless, boils at 54°, inflammable, burning with a green flame. Prepared by distillation from a mixture of strong aqueous muriatic acid and alcohol. HYDRIODIC and HYDROBROMIC ETHERS have been procured by heating the acids required with alcohol.

1460. CHLORIC ETHER is prepared by the action of chlorine on olefiant gas, and is commonly called Hydrocarburet of Chlorine. See 663, and also Chloral, 664.

1461. ACETIC, OXALIC, TARTARIC, CITRIC, and other ETHERS, have been obtained by distilling alcohol along with the acids required, or with mixtures by whose reactions the acids are liberated. They are usually regarded as compounds of an equivalent of acid and of sulphuric ether, part of the water of the alcohol being removed by the acid mixtures employed.

1462. OXAMID, or OXALAMID, is the name given to a compound which may be obtained by agitating oxalic ether and aqueous ammonia, or by distillation from oxalate of ammonia, which is decomposed; the term is derived from the first syllables of oxalic acid and ammonia. It is insoluble in cold water, and dissolved by boiling water. Exposed to heat alone it fuses, and is then partly sublimed and partly decomposed. Its symbol is $2\dot{}H+C^2N$. It may be regarded as a binaqueous cyanogen, or as a compound of binoxide of nitrogen and 2 of hydruret of carbon, as the following diagram shews :—

Fig. 65.

Materials.			Products.
2 eqs. water	Oxygen...	16	30 Binoxide of Nitrogen.
	Hydrogen	1	
	Hydrogen	1	
Byanogen...	Nitrogen	14	
	Carbon ...	6	7 Hydruret of Carbon.
	Carbon ...	6	7 Hydruret of Carbon.

Symb. $2\dot{}H+C^2N = \dot{\dot{}}N+2HC$.

1463. Again, oxamid and water can produce oxalate of ammonia $2\dot{}H+C^2N\&\dot{}H = \dot{\dot{\dot{}}}C^2+NH^3$.

1464. ETHERINE is the name now given to quadrocarbureted hydrogen (see 601) by many chemists, being regarded as the base of most ethereal compounds.

1465. ETHULE is the name given by Berzelius to a compound of hydrogen and etherine, which he considers the more immediate principle of ethers. It may be termed more specifically HYDRURET OF ETHERINE. Its symbol is H+4HC. Sulphuric ether may be regarded as an OXIDE OF ETHULE, this compound containing oxygen, carbon, and hydrogen, in the proportions in which they constitute

1 eq. of water and 4 of hydruret of carbon = HYDRATE OF ETHERINE, or sulphuric ether. The diagram represents this view more minutely.

Fig. 66.

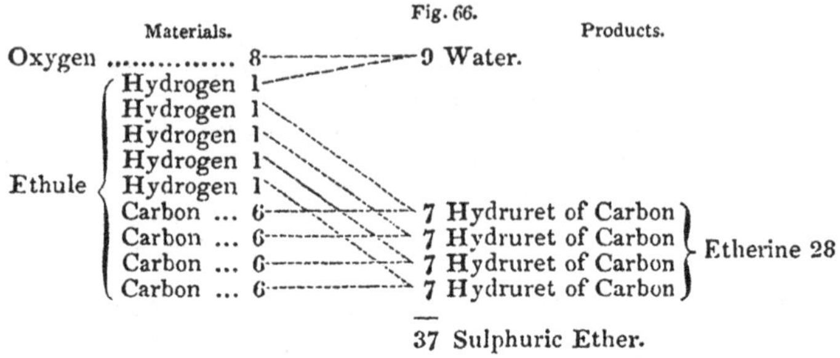

ALDEHYDE is a newly discovered colourless liquid, having a peculiar ethereal smell.—Sp. gr. 790.—Boils at 71°.—Symb. OC^2H^2.—It is termed aldehyde from the first syllables of alcohol and dehydrogenatus. Inflammable, burning with a blue flame, combines with water, alcohol, and ether.

1466. PREPARED by passing the vapour of ether through a long glass-tube heated to redness. The product being introduced into sulphuric ether, is retained in combination. Dry ammoniacal gas is then passed into the solution, which forms a crystalline compound with the aldehyde, termed AMMONIA-ALDEHYDE. From this compound it is procured as a hydrate by adding an equal weight of water, and then diluted sulphuric acid to unite with the ammonia, heating it afterwards in a retort. The product of distillation is the hydrated aldehyde, which is separated from the water by distilling it from chloride of calcium.

1467. Aldehyde may also be formed by the action of spongy platinum with air and alcohol, or by distillation from four parts of water and four of alcohol, mixed with six of binoxide of manganese and six of aqueous sulphuric acid.

1468. ALDEHYDIC ACID is prepared from aldehyde, and is composed of oxygen, hydrogen, and carbon, in the proportions represented in the following symbol: $O^2+H^2=C^4$. It may be regarded accordingly as acetic acid deprived of an equivalent of water.

1469. MERCAPTAN, or HYDRURET OF MERCAPTUM.—Symb. $H+S^2C^4H^5$. Liquid, colourless, volatile; dissolves sparingly in water, but freely in alcohol and ether.

1470. Procured by distillation from sulphovinate of baryta mixed with hydrosulphuret of baryta; it is associated at first with an ethereal fluid, from which it is separated by distillation. It acts energetically on the peroxide of mercury, producing 2 eqs. of water and 1 of bimercapturet of mercury. From this substance it is most easily prepared by the action of 2 eqs. of sulphureted hydrogen.

Fig. 67.

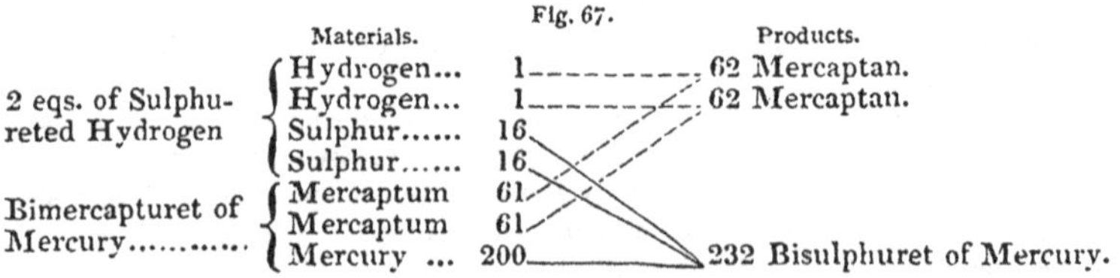

The name of this substance is derived from the powerful attraction which it exerts towards mercury.

1471. MERCAPTUM has not been procured in an insulated form. Its symbol, calculated from the composition of mercaptan, is $S^2C^4H^5$.

1472. XANTHOGEN* is the term applied by Zeise to the bisulphuret of carbon, which, when mixed intimately with alcohol and potassa, produces xanthate of potassa, the XANTHIC ACID being composed of alcohol and bisulphuret of carbon. Symbol of xanthic acid, $2S^2C+OC^2H^3$.

1473. A number of interesting galvanic experiments have lately been made by Mr Connell on alcohol and ether, from which he concludes that the elements of water exist as water in alcohol, but not in ether. Hydrogen gas was easily procured from alcohol after increasing its power of conducting electricity by 1–200dth part of potassa.

DIVISION II.—ANIMAL CHEMISTRY.

1474. The proximate principles of the animal kingdom, along with carbon, oxygen, and hydrogen, generally contain nitrogen, an element which renders them much more prone to decomposition than they otherwise could be, and is the cause of the more important peculiarities which they present when subjected to heat, or when undergoing putrefaction, a large quantity of ammonia being produced in both cases. Sulphur and phosphorus are also found in several proximate animal principles, though the quantity is often extremely small, producing sulphureted and phosphureted hydrogen during putrefaction. Several proximate animal principles have been observed to contain hydrogen in such proportions as to produce water with the oxygen and ammonia with the nitrogen. Food which has been cooked, and put into tin cases heated to 212°, filled completely with fluid, so that no air may remain in the interstices, may be preserved for years without acquiring any offensive taste or odour, putrefaction being entirely prevented.

1475. The principal products of PUTREFACTION are carbonic acid, carbureted hydrogen and ammonia, which are generally mixed with phosphureted and sulphureted hydrogen gases and a large quantity of water. The higher the temperature, if not excessive (beyond 90°), the more rapid the putrefaction; at a low temperature, it goes on slowly, and is arrested altogether at temperatures below 32°.

Animal matters are converted into a fatty compound called ADIPOCERE, by the long continued action of water, the air being excluded. Many regard it as an imperfect soap, composed of ammonia and fatty matter. Alcohol also produces a similar effect with fibrine.

CHAP. I. FIBRIN, ALBUMEN; GELATINE, OSMAZOME.

1476. These principles occur abundantly in the animal kingdom; the composition per cent. of the three first is very nearly the same, and is shewn in the following table—fractions have been omitted :—

	Carbon.	Hydrogen.	Nitrogen.	Oxygen.
FIBRINE, . .	53	7	19	19
ALBUMEN, . .	50	7	15	26
GELATINE, . .	47	7	16	27

1477. FIBRIN is the principal part of muscular fibre, and is found also in the

* This term is now applied by Dr Hope in the manner described at par. 1436.

blood of animals. Solid, tasteless, inodorous; has a whitish appearance; has some elasticity, and is rendered hard and brittle by drying. Soluble in strong ACETIC ACID, swelling at first, and forming a concentrated jelly. Decomposed by strong and by diluted NITRIC ACID, pure nitrogen being evolved from it when the acid is diluted; a yellow powder, called YELLOW ACID, is formed during the reaction of the nitric acid. Berzelius has affirmed that it is a compound of nitric acid and fibrine after it has been affected by the acid. With sulphuric acid, a solution is procured, containing a peculiar white matter called LEUCINE; the sulphuric acid is separated from it by chalk, the solution of the leucine being then filtered and evaporated. Diluted muriatic acid has little action on fibrine, and by the strong acid it is decomposed. Fibrine is also dissolved by concentrated solutions of potassa, soda, and ammonia, being at the same time decomposed. It is insoluble in water; alcohol converts it into a fatty matter.

1478. Procured from muscular fibre by macerating it in water, or by stirring newly drawn blood with a stick, when it collects in considerable quantity upon it.

1479. ALBUMEN is found abundantly in the solid form, and in solution in water, constituting in the latter case liquid albumen.

1480. SOLID ALBUMEN is found in the cellular membrane, and in a great number of other animal solids.

1481. LIQUID ALBUMEN forms the white of the egg, and almost the whole of the serum of the blood. It is a thick fluid, distinctly alkaline from the presence of soda, combines with cold water, and is coagulated at 160° by heat; it is also coagulated by alcohol, by sulphuric, nitric, muriatic, metaphosphoric, and many other acids; by ferrocyanate of potassa after the addition of acetic acid; by bichloride of mercury, muriates of tin and iron, acetate of lead, and by the infusion of galls. Phosphoric and pyrophosphoric acids do not precipitate it. The coagulated albumen generally carries along with it a portion of the precipitating agent. With bichloride of mercury, a precipitate of chloride of mercury and albumen is formed; or of the oxide of mercury, according to more recent investigations. Bichloride of mercury detects albumen in 2000 parts of water. An excess of albumen dissolves those precipitates which are compounds of albumen and an oxide.

1482. Albumen coagulated by heat, or by drying successive layers in the open air, resembles fibrine much, and can scarcely be distinguished from it by the action of tests. Berzelius states that it has no action on binoxide of nitrogen, but that fibrine produces a disengagement of oxygen.

1483. GELATINE is not found, like the preceding substances, in any animal fluids. Obtained principally from skin, bones, membranes, ligaments, and tendons. ISINGLASS is a purer variety, which is prepared from the sounds of the sturgeon and other fish. Solid, soluble in water, hot or cold; not coagulated by heat or acids; forms a solution which gelatinizes when cold, even when 100 parts of water are used with only 1 of gelatine. TANNIN precipitates it copiously; the compound is called TANNO-GELATINE, and is of the same nature with leather, which is usually prepared by the action of tannin (derived from oak-bark) with the skins of animals. GLUE consists of impure gelatine. It is insoluble in water; converted into a peculiar saccharine matter by sulphuric acid; not precipitated by bichloride of mercury or subacetate of lead.

1484. OSMAZOME is found associated with muscular fibre and other animal matters; it is particularly distinguished by its solubility in water and alcohol at any temperature, and by not forming a gelatinous solid when its solution is evaporated. Osmazome is regarded as the matter which gives to broth its peculiar flavour.

CHAP. II. BONE, MUSCLE, &c.

1485. **Bones** contain about 33 per cent. of animal matter, and 67 of earthy substances. The animal matter is composed principally of gelatine and marrow or fatty matter. The following are the component parts of the earthy matter in 100 parts of bones, omitting fractions:—

Phosphate of lime, about	51 parts.
Carbonate of lime,	11 ...
Fluoride of calcium,	2 ...
Phosphate of magnesia,	1 ...

Soda, chloride of sodium, and water in smaller proportion.
Silica and alumina, with
Oxides of iron and manganese, have also been detected.

1486. Exposed to heat in the open air, the animal matter is consumed, and the earthy substances alone left. Exposed to heat without access of air, ammonia, inflammable gases, oily matter, water, and other substances, are evolved, much of the carbon of the decomposed animal matter remaining with the earthy substances of the bone. In this condition it is termed Ivory Black, which is much employed as a decolorizing agent, charcoal from animal substances (usually called Animal Charcoal) being very powerful in this respect. If the charcoal be required perfectly free from earthy matter, muriatic acid may be employed to dissolve it; and when it has been removed by solution, the remaining charcoal should be well washed, and heated to redness, before it is used to destroy animal or vegetable colouring matter.

1487. If bones be kept for some time in diluted muriatic acid, all earthy matter is removed, and the animal matter which remains retains the original form of the bone.

1488. Teeth are composed of the same materials as bones, but contain less animal matter.

1489. Horns, hoofs, nails, tendons, the cuticle, and the true skin, are composed principally of gelatine; horns contain also coagulated albumen, and a portion of earthy matter.

1490. The muscles are composed principally of fibrine, with albumen, gelatine, osmazome, fatty, and saline matter.

1491. Hair, wool, and feathers, are considered to contain a peculiar animal matter. Silica, sulphur, iron, manganese, and other substances, more particularly salts of lime, have also been detected in them.

1492. In brain and the matter of the nerves, 80 per cent. of water are found. Albumen, fatty matter, and osmazome, constitute the other principal ingredients. A variable proportion of phosphorus has also been detected, along with minute quantities of salts and sulphur.

CHAP. III. BLOOD, RESPIRATION, ANIMAL HEAT.

1493. The blood is a fluid slightly saline, unctuous, and has a peculiar odour. Sp. gr. 105, and temperature above 97° when newly drawn, or while circulating in the bloodvessels; it appears to be homogeneous, but by the microscope it is found to consist of a fluid almost without colour, in which numerous red particles are suspended.

1494. When removed from the bloodvessels, a halitus or vapour arises from the surface, composed of water and a little animal matter, and after a few minutes the whole mass gradually assumes a solid consistence. Shortly afterwards a few drops of yellowish fluid gather on the top, and, finally, the blood spontaneously separates into two parts, the CLOT or CRASSAMENTUM, which is thick and solid, and the SERUM or fluid portion. From 2 to 3 parts of crassamentum are usually procured, with 1 of serum.

1495. The conversion of the fluid mass into the solid form is called the COAGULATION of the blood, and it commences within two or three minutes after its removal from the bloodvessels; the clot or coagulum, however, often continues to contract slightly for two or three days; it then assumes the form of a cup, and floats amidst the serum. The cause of the coagulation is not known; it has been attributed to a vital action, the blood being considered to have the property of vitality as well as the living solids. It indeed contains organized solids floating in a transparent medium.

1496. The coagulation is accelerated by exposing the blood to a temperature of 120°, or drawing it from a small orifice into a shallow vessel.

1497. It coagulates quickly if the air be rapidly exhausted from the vessel in which it is received; and it has been observed to coagulate speedily in proportion to the depression of the vital energies, as, for instance, in hœmorrhage. Hence the blood last removed generally coagulates first. Alum, and the sulphates of zinc and copper, promote this change. The tint of the coagulum is much affected by the colour of the vessel in which the blood is received, if it be green, as Dr Newbigging pointed out.

1498. Saturated solutions of muriate of soda, muriate of ammonia, nitrate of potassa, and potassa, death arising from violent mental emotions, or preceded by severe exercise, prevent the process of coagulation. Low temperatures produce a similar effect, or retard it much; thus, blood which coagulates in five minutes at 60°, requires fully an hour at 40°.

1499. It has been stated that the blood does not coagulate in cases of death induced by lightning, but this has lately been contradicted. In animals killed by a powerful galvanic battery the blood has been found coagulated.

1500. Besides a particular exhalation from the blood, heat is evolved during coagulation. Carbonic acid gas was supposed to be disengaged; but it is not now considered that any of this gas is evolved.

1501. The blood, according to M. Le Canu, consists of the following substances, in 1000 parts:—

Water,	785.590
Fibrin,	3.565
Albumen,	69.415
Colouring matter,	119.626
Crystalline fatty matter, termed Seroline, (Cholesterine ?)	4.300
Oily matter,	2.270
Extractive soluble both in alcohol and water,	1.920
Albumen combined with soda,	2.010
Chlorides of sodium and potassium, with phosphates, sulphates, carbonates, of potassa and soda,	7.304
Carbonates of lime and magnesia; phosphates of lime, magnesia, and iron; peroxide of iron,	1.414
Loss,	2.586

1502. Small portions of alumina, silica, and manganese, have been found in the blood, and even a minute trace of copper, by M. Sarzeau and Dr O'Shaughnessy.

1503. Baruel maintains that the blood contains, in addition to the preceding principles, a volatile matter peculiar in each species, which is disengaged when the blood is mixed with strong sulphuric acid.

1504. The proportion of the different substances in blood varies at different periods of life, in different individuals, and in disease. The proportion also of the serum to the clot varies much from the shape of the vessel in which the fluid is received. The fatty matter has been regarded as Cholesterine.

1505. From experiments lately made on the changes produced in the composition of the blood by repeated bleedings by Mr Andrews, it appears that the albumen and salts decrease at each bleeding; the diminution is, however, very variable, and, even after the fourth time, does not amount to one and a half per cent. In the globules the same diminution takes place, but to such a degree that they are at least reduced to less than one-half their original quantity.

1506. The proportion of solid matter of the serum, and solid matter of the clot, is variously estimated, but Prevost and Dumas give the following relative quantities, in 1000 parts of human blood:—

Water,	784
Solid matter of crassamentum,	129
.... serum,	87

1507. In the Crassamentum the principal solids are the fibrine and colouring matter of the blood, mixed with albumen derived from the serum. By washing in a cloth with water, all the colouring matter may be removed, the fibrine being left. The fibrine is found not only in the red globules, but also in solution in the serum, as it circulates in the living system.

1508. Colouring matter of the Blood.—Regarded formerly as depending essentially upon iron for its tint, which is attributed now to a peculiar animal matter resembling albumen, and called Hematosine. It differs from albumen in its colour, and is black when pure; it has a reddish colour when reduced to powder. It is more easily coagulated by heat than albumen, and is not precipitated by the acetate or subacetate of lead. It contains carbon, oxygen, hydrogen, and nitrogen, with a minute quantity of iron. It acts with other agents in the same manner as albumen.

1509. When chlorine is transmitted through a solution of the colouring matter, a white flocculent matter is precipitated, and a transparent fluid is obtained, in which the iron may be detected by all the usual tests. Iron cannot be detected by the usual reagents, when dissolved in a solution containing organic matter.

1510. Prepared by diluting a solution of colouring matter in albumen with 10 parts of water, and heating the liquid, when the colouring matter is separated by coagulation at the temperature of 149°, while albumen remains in solution till heated to 160°. It is also precipitated by several metallic oxides. A solution of the colouring matter in excess may be procured by stirring the clot in water, having drained it previously on bibulous paper, after cutting it in thin slices. The solution of colouring matter in albumen is procured by stirring newly drawn blood, so as to remove the fibrine.

1511. Erithrogen (from ἐρυθρος, ruber) is a term applied by Bizio to a peculiar animal principle obtained by him in a diseased gall-bladder, and which he considered as the base of the colouring matter of the blood. It is turned red by nitrogen. See 1438, where this term is used in another sense.

1512. The serum constitutes the fluid portion of the blood; it is of a pale yellow colour, with a slight tinge of green, and sometimes presents a milky appearance. Sp. gr. 1.030. It contains free alkali (soda).

1513. Coagulated by heat, acids, alcohol, and by galvanism. On cutting and pressing the coagulum when produced by heat, a small quantity of colourless limpid fluid exudes, called the SEROSITY, containing a considerable portion of the saline matter of the blood, and also a portion of animal matter.

1514. According to Marcet, 1000 parts of the serum consist of—

Water, ...	900.
Albumen, . . .	86.8
Muriate of potassa and soda, . .	6.6
Muco-extractive matter, . .	4.
Carbonate of soda, . . .	1.65
Sulphate of potassa, . .	0.35
Earthy phosphates, . . .	0.60

1515. RESPIRATION consists in the inspiration and expiration of air, during which the air received into the lungs meets with the blood, when it changes from the dark purple colour of venous blood to the bright and brilliant red colour which it presents in the arteries. No difficulty is now entertained with respect to the air penetrating through the thin membrane of the cells of the lungs, as numerous experiments, particularly those of Dr Mitchell and Dr Faust, have shewn that air can pass through membranous matter, and affect chemically the contents within.

The experiments on the diffusion of gases illustrate the passage of air through apertures impervious to water; while the movements that take place in different fluids separated by a membranous partition, also clearly prove the facility with which an interchange of principles can ensue with great force where it was not previously suspected. Dutrochet, who made many interesting experiments on this subject, found that a bladder filled with a syrupy fluid and placed in water, soon absorbed so much of the water that it burst, a portion of the viscid fluid also escaping through the pores. ENDOSMOSE is the term applied to this peculiar action as it is observed in liquids, and EXOSMOSE to the passage of a portion of fluid from the interior to the other portion of liquid with which it may be surrounded; this exosmic movement always accompanies the endosmic action. The extensive surface on which the fluid is spread in the cells of the lungs, must be peculiarly favourable for the absorption of oxygen from the air by the blood, and the evolution of carbonic acid.

1516. Blood agitated with air or oxygen becomes of a florid red in the same manner as in the lungs; but with nitrogen and with carbonic acid the colour is darkened. The quantity of air affected appears to correspond with the amount of colouring matter in the blood. The presence of saline matter, as in the serum of the blood, is essential to the change of colour; it does not take place without it, however freely the air or oxygen may be supplied, as Dr Stevens proved. The experiments of Dr Gregory and Mr Irvine have shewn that oxygen is necessary to induce the red tint in the globules diffused through serum, or any similarly diluted solution of saline matter, though the change may be produced in a strong saline solution without any oxygen. ARTERIALIZATION is the term applied to the changes that are produced in the fluid derived from the food, as it is converted into blood.

157. During respiration, the quantity of oxygen in the air is diminished, and in man it is replaced by an equal bulk of carbonic acid gas; in other animals, the quantity of this gas given out is occasionally observed to be greater, and sometimes less than the oxygen consumed. Every minute, it has been calculated by Messrs Allen and Pepys, 26 c. i. of carbonic acid are produced, an estimate considered rather high by many chemists; the air given out from the lungs contains, ac-

cording to other estimates, 3.6 per cent. of carbonic acid; according to them, from 6 to 8 per cent. of this gas.

1518. According to Dr Thomson, "there are 20 respirations per minute, and 16 cubic inches of air taken in and thrown out at each respiration; we have, therefore, 28.800 respirations in 24 hours, and 460.800 cubic inches of air passing through the lungs. Of this $\frac{3.72}{100}$ or 17,141.76 cubic inches are converted into carbonic acid gas. Now, 100 cubic inches of carbonic acid weigh nearly 50 grains; so that the weight of carbonic acid formed is 8,570.8 grains, $\frac{3}{11}$ths of which, or 2337.5 grains, are carbon. This amounts to nearly nine ounces avoirdupois, or somewhat more than half a pound."—*Records of Science.*

1519. Dr Thomson's, Dr Prout's, and Dr Fyfe's experiments, shew that the quantity of carbonic acid evolved at different temperatures, varies much under different circumstances, and even at different periods of the day.

1520. By a forced expiration, about 200 c. i. of air may, on an average, be expelled from the lungs.

1521. The nitrogen of the air is little affected, apparently, by respiration; occasionally its quantity appears to be increased, and sometimes it is diminished, the effect varying with the seasons and other circumstances.

1522. Animal Heat.—The discovery of carbonic acid in the air disengaged from the lungs during respiration, was made by Dr Black. He considered respiration analogous to combustion, and that the carbonic acid is formed in the lungs. Crawford, adopting his views, believed that the capacity of the blood for caloric is increased at the moment the carbonic acid is produced, and hence the reason why no burning heat is perceived in the lungs; but the capacity of the blood, he supposed, is diminished as it passes from arterial to venous blood in the extreme capillaries, when the heat that had originally been produced (though not rendered sensible in the lungs) is evolved, diffusing an equal degree of warmth over the whole body. His experiments, however, as to the relative capacities of oxygen, carbonic acid, venous and arterial blood, on which the theory rests, have not been supported by other chemists.

1523. Mr Ellis considered that carbon is separated from the blood as an excreted product, and then acts on the air inspired.

1524. Hassenfratz and Le Grange proposed another view of the manner in which the carbonic acid is produced, and it is most generally received at present. They considered that the oxygen of the air is absorbed by the blood, and a corresponding quantity of carbonic acid evolved, produced during the course of the circulation by the oxygen which had been previously absorbed. Carbonic acid gas has been detected in venous blood, being evolved when it is transferred directly from the living body into an atmosphere of hydrogen gas.

1525. The skin affects the air much in the same manner as the lungs, carbonic acid being produced and oxygen consumed.

1526. In some animals, respiration is carried on entirely by the skin, and a considerable quantity of carbonic acid evolved.

1527. The production of animal heat was considered by Dr Black to depend upon the formation of carbonic acid by the oxygen of the air combining with the carbon of the blood. Numerous experiments have now proved, that the greater the heat produced in the body, the greater the consumption of oxygen in the lungs; it is also supposed that this operation is not the only source of animal heat, but that it may be developed in part by other operations going on at the same time.

1528. By disease, blood is much altered in its properties. In cases of cholera it is very much affected; its colour becomes dark, sometimes it acquires the consistence of tar, and is less readily affected by the oxygen of the air. It loses much water, and most of its saline matter, the proportion of albumen and colouring matter being increased. Its density is greater, and it does not coagulate.

1529. Blood occasionally presents a white appearance, owing to the presence of fatty matter in considerable quantity, which is detected by ether dissolving it, and giving a solution, from which it may be procured by evaporation.

1530. In cases of inflammatory action, the crassamentum is covered with a coat of pure fibrine, usually called the BUFFY COAT. This arises from the blood being so altered in its qualities, that the fibrine it contains in solution coagulates more slowly than the rest of the blood, and part of it is deposited above the red clot. The red globules of the blood are considered heavier than pure fibrine, consisting of a small portion of colourless fibrine in the centre, which is surrounded by the colouring matter of the blood. When the blood is removed from the body, and the colouring matter escapes from the globule, the fibrine from the centre adheres firmly together.

1531. The blood is affected to a great extent in a number of other diseases, though this may not in general be so easily recognised as in the preceding cases, chemical analysis being required to point out the change. Occasionally, however, the change is sufficiently evident, as in jaundice, when the blood acquires a greenish-yellow tint in consequence of the absorption of bile. The BLACK VOMIT observed in yellow fever is regarded as a compound of blood and muriatic acid. Urea is frequently observed in the blood, more especially in those cases when the secretion of urine is suppressed.

CHAP. IV. SALIVARY, PANCREATIC, AND GASTRIC JUICES—DIGESTION—BILE.

1532. The SALIVA contains a small quantity of solids in solution, scarcely amounting to 1 per cent. The solid matter is composed of a peculiar animal matter and saline substances, among which free soda and sulphocyanate of potassa have been detected. It varies, however, in its composition, and has been frequently observed acid, neutral, and alkaline.

1533. PANCREATIC JUICE.—Regarded formerly as being of the same nature with saliva, though now considered very different, containing a little albumen, curdy matter, osmazome, a free acid (acetic ?), but no sulphocyanic acid is present.

1534. GASTRIC JUICE.—This fluid is secreted in its proper form only from the stimulus of food, when muriatic acid may be distinctly traced in it, to which the great solving powers which it possesses are attributed; acetic acid is also associated with it. The muriatic acid is probably derived from common salt, and to the soda produced, as the muriatic acid is removed, the alkaline reaction of the blood may perhaps be attributed. The stomach itself is supposed to be defended from the action of the corrosive acid by assuming a peculiar electric condition. In cases of sudden death, the stomach is often found corroded in consequence of the action of the acid on its fibres. Gastric Juice acts powerfully in coagulating milk.

1535. The gastric juice acting on the food produces a pulpy mass, termed chyme, from which, in the intestines, a milky fluid, THE CHYLE, is absorbed; this contains the nutritious matter derived from the food, and is conveyed to the heart, and thence to the lungs, where it acts with the air, and is converted into arterial blood.

1536. The insipid fluid formerly described by chemists as gastric juice, must be regarded perhaps as merely a portion of the salivary and pancreatic juices.

1537. Bile is a greenish-yellow coloured fluid, generally rather viscid, having a sweetish bitter taste and nauseous odour. Heavier than water, alkaline; coagulated by acids.

Thenard regards the bile of the ox as a compound of about 7 parts of water and 1 of animal and saline matter, composed of—

Picromel.	Muriate of soda.
Resin.	Muriate of potassa.
Yellow matter.	Sulphate of soda.
Soda.	Phosphate of lime.
Phosphate of soda.	Magnesia and oxide of iron.

The saline matter constitutes a small proportion of the ingredients.

Cholesterine, an odoriferous animal matter, and another peculiar animal matter, osmazome, gluten, cholic acid, and some fatty substances, have also been found in bile. In human bile, similar ingredients have been detected.

1538. Picromel.—Solid, crystalline, soluble in alcohol and water; taste sweet. Prepared from bile by precipitating sulphuric acid and some other substances by acetate of lead, then adding subacetate of lead, the oxide falling down with the picromel and resin. By hydrosulphuric acid acting on the precipitate suspended in water, sulphuret of lead is formed, being left undissolved along with the resin; the picromel remains in solution.

1539. Cholic Acid is solid, crystalline, reddens litmus, and has a sweet taste.

1540. Biliary Calculi are composed principally of cholesterine, and the colouring matter of the bile. Sometimes they contain no cholesterine.

1541. Cholesterine.—White, crystalline, with a pearly lustre. Melts at 278°; does not form a soap with potassa. Insoluble in water; dissolved abundantly by boiling alcohol; sparingly soluble in cold alcohol. By the action of nitric acid, cholesteric acid is produced.

CHAP. V. MILK AND CHYLE.

1542. Milk contains the following substances, of which the first, water, constitutes nearly 929 parts in 1000:—

Water.	Muriate of potassa.
Butter.	Acetate of potassa.
Caseous matter.	Phosphate of potassa.
Sugar of milk.	Phosphate of lime.
Lactic acid.	Traces of iron.

1543. Cream contains rather more than 3 per cent. of caseous matter, and 4 of butter, the rest being whey.

1544. Whey consists principally of water, with small portions of animal matter, and a large quantity of a peculiar saccharine matter, called Sugar of Milk, which may be procured by evaporation.

1545. Butyrine is the name given to oily matters which constitutes butter.

1546. Caseous Matter is the curdy substance obtained from milk coagulated by rennet, the infusion made by the action of water upon a portion of the stomach of the calf, which is powerful in coagulating milk. It always contains in this condition some foreign matter associated with it, being soluble in water when pure, and forming a mucilaginous solution. Sulphuric, nitric, muriatic, and other acids; alcohol, the infusion of galls, and a variety of other substances, coagulate milk by combining with the caseous matter.

1547. Caseous matter is maintained by some chemists to contain two distinct principles, CASEIC ACID, and CASEOUS OXIDE or APOSEPIDINE. Others again regard it as approaching very nearly to coagulated albumen in its leading characters.

1548. CHYLE is the milky looking fluid taken up from the chyme. It approaches in its characters to blood, but has only a slight pink tint, and contains less solid matter. It forms a less firm crassamentum during coagulation, and from its serum a flocculent precipitate is obtained by heat, termed by Prout INCIPIENT ALBUMEN. The chyle of two dogs analyzed by him contained from 89 to 94 of water, the rest being fibrine, incipient albumen, albumen with a slight pink tint, and minute quantities of sugar, and oily and saline matters.

CHAP. VI. OLEAGINOUS AND FATTY SUBSTANCES.

1549. These resemble much in all their leading characters the fixed oils of vegetables. STEARINE, a principle analogous to the margarine of fixed vegetable oils, is found in most of them associated with variable proportions of OLEINE. Berard prepared a substance very similar to fat, by passing through a red hot tube a mixture of carbonic acid, carbureted hydrogen, and hydrogen. Dobereiner succeeded in producing an analogous compound with coal gas and watery vapour.

1550. STEARINE is obtained with facility in brilliant crystals when deposited from a hot ethereal solution. It is very soluble in hot ether, sparing, soluble in cold ether. It is also soluble in boiling alcohol. Melts at 129. PREPARED by boiling mutton suet in ether, after melting it to separate any membranous matter, and removing the adhering solution from the crystals by bibulous paper; this process is repeated with the crystals several times. Similar processes may be adopted in preparing stearine from other fatty matters.

1551. When boiled with a solution of potassa or soda, it is resolved into STEARIC ACID and glycerine. The stearic acid may be separated by neutralizing the alkali with sulphuric acid. See 1354.

1552. MARGARINE is the name given to another fatty matter very similar to stearine, but more soluble in ether, and melting at 117°. It is procured by allowing the matter separated from the stearine (see 1550) to evaporate and crystallize spontaneously. See 1355.

1553. Olein is obtained by pressing lard in bibulous paper, to which it adheres. It is similar to that procured from vegetable substances.

1554. ADIPOCIRE. See 1475.

1555. AMBERGRIS is considered to be a concretion produced in the stomach of the spermaceti whale. It is found floating on the sea coast of India and Africa. It consists principally of a peculiar fatty matter, called Ambreine, which resembles cholesterine.

1556. BUTYRINE is the fatty matter that produces butter.

1557. CHOLESTERINE. See Bile.

1558. DIPPEL'S OIL is the name given to a thin limpid oil, the product of the destructive distillation of animal substances.

1559. FAT, HOGSLARD, and SUET, are compounds of stearine and oleine in various proportions; they melt at various temperatures between 59° and 102°. The stearine and oleine differ often in the fat obtained from different animals.

1560. HIRCINE is procured from the fat of the goat and sheep.

1561. SPERMACETI is prepared from the fatty matter found in the head of the spermaceti whale. Solid, white, crystalline, insoluble in water, soluble in ether and alcohol. Melts at a temperature below 212°. It is usually mixed with a little fluid oil, and is termed CETINE when purified by solution in boiling alcohol and crystallization. ETHAL is a solid fatty matter which remains after the separation of margaric and oleic acids; boiling cetine with potassa or soda, so as to produce soap.

1562. SPERMACETI OIL is the fluid expressed from the fatty matter from which the spermaceti is obtained.

1563. TRAIN OIL is procured by heating blubber to 212°. Its offensive odour arises from decomposed animal matters which are mixed with it.

CHAP. VII. MUCUS, PUS, &c. LACTIC AND FORMIC ACIDS.

1564. MUCUS.—The existence of a distinct principle to which this name has been applied is doubtful. The mucus described by Dr Bostock is soluble in hot and cold water, and does not gelatinize. Tannin and bichloride of mercury do not precipitate it. The mucus of the nose is rendered transparent by water, but not dissolved. It is dissolved by nitric acid, dilute sulphuric acid, and potassa.

1565. PUS varies much in its qualities, according to the nature of the source from which it is produced. Healthy pus is a bland, thick fluid, apparently homogeneous, but composed of a thin transparent fluid, with opaque globules floating in it. Sp. gr. 1.030. Neutral, but becomes acid by the action of the air. Soluble in sulphuric, nitric, and muriatic acids, and in alkalis. Ammonia produces a gelatinous mass with it.

The following are the principal tests which have been proposed for distinguishing pus from mucus:—

TESTS.	MUCUS.	PUS.
Mixed with an equal weight of water, and then with an equal weight of a saturated solution of carbonate of potass,	does not gelatinize,	produces a jelly.
Diffused through water,	from a catarrh, it floats.	precipitated.
Dissolved in potassa, and water added,	not affected.	precipitated.
Dissolved in sulphuric acid, and water added.	remains suspended in the water.	precipitated.

1566. FLUID OF SEROUS SURFACES.—Composed principally of water, with small portions of albumen, mucus, and saline matter. The LYMPH which lubricates the cellular membrane is considered of analogous composition. Small portions of LACTIC ACID are also found in it.

1567. LACTIC ACID has been found in most animal fluids, and in a number of vegetables; it was first obtained from sour milk, from which its name is derived. Its concentrated solution is syrupy, very acid, and can displace acetic acid from its combinations. It is prepared by evaporating solutions containing it to a syrupy consistence, extracting the lactic acid by alcohol. By combination with oxide of zinc, separating it afterwards by baryta, and ultimately removing the baryta by sulphuric acid, it is obtained in a pure form.

1568. Formic Acid.—Symb. HO^3C^2 = hydrogen 1 + oxygen 24 + carbon 12. —Eq. 37.—The name given to the acid obtained from the acid liquid of ants. It may be formed artificially by distillation from tartaric acid mixed with diluted sulphuric acid and peroxide of manganese. It will be seen from the annexed diagram that it may be regarded as a compound of 2 eqs. of carbonic oxide and 1 of water, into which it is easily resolved by heating it cautiously with strong sulphuric acid.

Fig. 68.

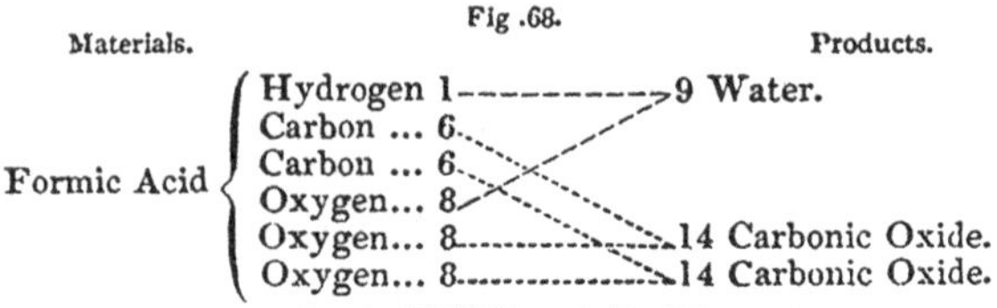

Symb. $O^3C^2H = 2\cdot C + \cdot H$.

CHAP. VIII. UREA, URIC ACID, PURPURIC ACID, CYANURIC ACID, URINE, URINARY CALCULI.

1569. Urea.—Symb. $2H^2OCN$ = hydrogen 4 + oxygen 16 + carbon 12 + nitrogen 28.—Eq. 60.—Solid; crystallizes in 4-sided prisms. Sp. gr. 1.35. Very soluble in water and alcohol; is not acid nor alkaline. Decomposed by heat, producing carbonate of ammonia and cyanuric acid. Decomposed quickly as it exists in urine, but its solution in water, when pure, is not speedily decomposed.

1570. With nitric acid it forms a sparingly soluble crystalline compound; this acid is accordingly employed to detect urea.

1571. Preparation.—Evaporate urine to the consistence of a syrup. When cold, add gradually nitric acid (free from nitrous) till a dark crystallized mass is formed, composed principally of nitric acid and urea. Wash it repeatedly with cold water, and dry by pressure between folds of bibulous paper. Dissolve carbonate of potassa in 3 parts of water; add the solution till the nitric acid is neutralized; evaporate the solution that the nitrate of potassa may crystallize on cooling. Mix the drainings from the crystals which contain urea in solution, with animal charcoal, so as to produce a thick paste. After 6 or 8 hours, add cold water to extract the urea, thus freed from colouring matter; filter and evaporate the solution to dryness. Boil the urea in alcohol, then filter and crystallize.

1572. Urea is formed also when water is heated with cyanate of ammonia; 1 eq. of water with 1 of cyanate of ammonia, containing the same elements in the same proportion. The diagram shews the relation that urea and the aqueous cyanate of ammonia bear to each other.

Fig. 69.

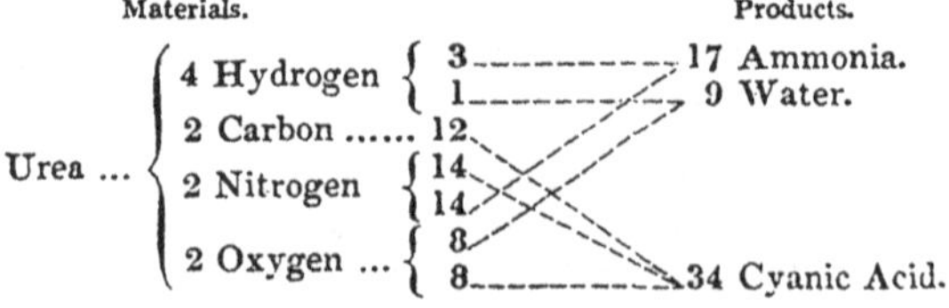

Symb. $2H^2CNO = \cdot H + \cdot C^2N + NH^3$.

The aqueous cyanate of ammonia is supposed to be in solution.

1573. By examining the changes symbolically, and with the aid of a diagram, the production of carbonate of ammonia by the action of water with urea may be easily seen. The 2 eqs. of carbonate are supposed to be in solution.

Fig. 70.

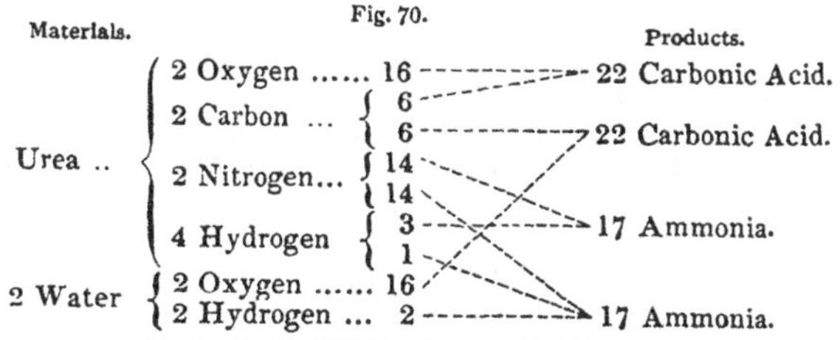

Symb. $2{\cdot}H\&2H^2CNO = 2(:C+NH^3)$.

1574. Uric or Lithic Acid.—Symb. $H^2O^3C^6N^2$ = hydrogen 2 + oxygen 24 + carbon 36 + nitrogen 28.—Eq. 90.—White when pure, but generally of a brownish colour; tasteless, inodorous; insoluble in alcohol; 10,000 parts of cold water are required for its solution. Reddens litmus feebly; expelled from its salts by carbonic acid, but, if boiled with carbonates, it expels carbonic acid. Particularly distinguished by the crimson tint which it produces when heated with nitric acid, purpurate of ammonia being formed. Decomposed by heat, cyanuric acid is produced.

1575. Prepared from calculi of uric acid, or from the uric acid deposited from acidulated urine, by dissolving it in a solution of potassa, and adding an acid to precipitate it from the urate of potassa.

1576. Purpuric Acid is white when pure, and is particularly distinguished by the brilliant coloured purple compounds which it forms with several of the salifiable bases. Formed in combination with ammonia by the action of nitric and uric acids. The ammonia may be displaced by potassa, and the purpuric acid precipitated by adding sulphuric acid to combine with the potassa. The erythric acid of Brugnatelli, and the sediment often deposited from urine in fevers, and called at one time Rosacic Acid, are considered by Prout to be composed of purpurate of ammonia.

1577. Cyanuric Acid, called also Pyrouric Acid, is formed when uric acid is heated, or by the action of chlorine on different compounds containing cyanogen and water. Urea also may be made to produce this acid.

1578. A peculiar colouring matter, not containing any purpuric acid, has also been discovered in the urine.

1579. Urine is a transparent limpid fluid, of an amber colour; sp. gr. 1.0224, when recently discharged it has an acid reaction, but after a short time it acquires decided alkaline properties. The following are the component parts of urine, according to Berzelius, in 1000 parts:—

Water,	933.00
Urea,	30.10
Uric Acid,	1.00
Free lactic acid, and lactate of ammonia with animal matter,	17.14
Mucus of the bladder,	0.32
Sulphate of potassa,	3.71
.... soda,	3.16
Phosphate of soda,	2.94

Phosphate of ammonia, . .	1.65
Muriate of soda, . . .	4.45
.... ammonia, . .	1.50
Earthy matters, with a trace of fluate of lime,	1.00
Siliceous earth, . . .	0.03

Sulphur, phosphorus, and albumen, are also found, but in very small quantities. In children, and also in graminivorous animals, a considerable amount of benzoic acid may be detected. Its specific gravity varies very much, both in health and disease.

1580. Urine is quickly decomposed spontaneously; and as the urea is resolved into carbonate of ammonia, phosphate of lime and phosphate of ammonia and magnesia are deposited.

1581. From disease the urine is often much changed in its qualities; the following are the principal alterations.

1. *Deposition of Calculi, &c.*

1582. The urine often becomes so loaded with different materials, that much is deposited in the solid form before it is discharged, giving rise to the production of URINARY SAND or CALCULI, according to the cohesion of the precipitated matter.

1583. URIC ACID CALCULI are of a brownish-yellow colour, and generally consist of different layers of acid. They are decomposed by heat, soluble in potassa, produce purpurate of ammonia by nitric acid; see 1574. In most calculi, small portions of uric acid may be detected. An excess of uric acid, or the decomposition of urate of ammonia by other acids, are considered the principal causes of the deposition of uric acid.

1584. URATE OF AMMONIA CALCULI have a clay colour; evolve ammonia when heated with potassa. With the other agents mentioned in the preceding paragraph, the same phenomena are produced as with uric acid calculi.

1585. OXALATE OF LIME CALCULI are rough and tuberculated on the surface. Heated to dull redness they produce carbonate of lime. Heated to whiteness nothing is left but quicklime. With sulphuric acid, sulphate of lime is formed, and then the oxalic acid may be separated in solution by water.

1586. PHOSPHATE OF LIME CALCULI.—Not decomposed by heat; insoluble in potassa; soluble in diluted nitric or muriatic acid; give no ammonia when heated with potassa; not dissolved by cold acetic acid.

1587. PHOSPHATE OF AMMONIA AND MAGNESIA CALCULI evolve ammonia when heated alone, or with potassa. Not dissipated by heat, though the ammonia is expelled. Soluble in diluted nitric and muriatic acids; soluble also in cold acetic acid.

1588. The FUSIBLE CALCULUS is a mixture of phosphate of lime with phosphate of ammonia and magnesia. It is melted by heat. Cold acetic acid dissolves the phosphate of ammonia and magnesia, but does not affect the phosphate of lime.

1589. The CARBONATE OF LIME CALCULUS is distinguished in the same manner as common carbonate of lime. A portion of animal matter is generally blended with it. A calculus composed of oxalate and carbonate of lime has lately been noticed. Both these varieties, however, are extremely rare.

1590. The ALTERNATING CALCULUS consists of alternate layers of some of the preceding calculi.

1591. SILICEOUS GRAVEL has occasionally been noticed in some urinary complaints. It is not affected by heat, is insoluble in acids, fuses with alkalis added in excess, and produces silicated potassa.

1592. CYSTIC OXIDE CALCULI contain a peculiar animal matter, cystic oxide, which is soluble in acids, alkalis, alkaline carbonates, and lime-water.

1593. XANTHIC OXIDE CALCULI consist of another peculiar animal matter. With nitric acid it produces a lemon yellow coloured compound.

1594. Fibrinous calculi are composed of fibrine. The last three varieties of calculi are extremely rare, and are decomposed by heat, in the same manner as other animal substances.

1595. The uric acid, and the phosphate of ammonia and magnesia calculi, are those most frequently observed.

2. *Production of Sugar, Albumen, &c.*

1596. SUGAR is found in considerable quantity in the urine of individuals affected with diabetes; 6 per cent. of sugar may often be procured from it. Professor Kane obtained a still larger quantity.

1597. ALBUMEN is often found in large quantity in the urine of individuals affected with some varieties of dropsy, coagulating when exposed to heat like the serum of the blood. In some cases it has coagulated even within the bladder.

1598. UREA is sometimes found in excess in urine. Dr Prout states, that, when this is the case, nitric acid added in an equal bulk to a few drops of urine in a watch-glass, produces a crystalline precipitate of nitrated urea in half an hour. Healthy urine produces it more slowly. It is not absent in diabetic urine, as was at one time supposed.

1599. In some diseases of the liver, the urine becomes tinged with bile, and has a deeper yellowish tint than usual. Muriatic acid produces a green tint in urine charged with bile.

APPENDIX.

MINERAL WATERS.

1600. Mineral waters are usually divided into four classes, Acidulous, Sulphureous, Saline, and Chalybeate. The different ingredients may be detected by the various tests described in the preceding pages.

1601. ACIDULOUS MINERAL WATERS owe their peculiar qualities to carbonic acid, which escapes by heat, or on exposure to the air for some time. They sparkle when poured from one vessel into another, and are generally found with various proportions of carbonates of lime, iron, and other saline matters in solution. CARLSBAD, SELTZER, SPA, and PYRMONT, are the most noted acidulous waters. The carbonates are retained in solution by excess of acid.

1602. SULPHUREOUS MINERAL WATERS contain sulphureted hydrogen, to which they owe their offensive taste and odour, as those of HARROWGATE and MOFFAT. The water of St Bernard's Well, in the neighbourhood of Edinburgh, contains it in small quantity.

1603. SALINE MINERAL WATERS generally contain various proportions of sulphates and muriates of soda, lime, and magnesia. Potassa, hydriodic acid, hydrobromic acid, a free alkali, and siliceous matter, are also found in some saline mineral waters. SEA WATER is usually classed along with the saline mineral waters; its principal saline ingredient is common salt, of which the water of the Frith of Forth contains about 1-45th; sulphate of lime, sulphate of magnesia, and muriate of magnesia, are present in smaller quantity. Iodine, bromine, and potassa have also been detected in it.

1605. CHALYBEATE MINERAL WATERS contain iron, oxidated and combined with sulphuric or carbonic acid. By boiling or exposure to the air, carbonic acid escapes, when the iron is carbonated and oxide of iron is precipitated. An ochry deposit also takes place from the sulphate as the iron becomes peroxide by the action of the oxygen of the air. The strongest chalybeate water known is that which is obtained from Dollar. It has a deep red colour, and contains peroxide of iron in combination with sulphuric acid and alumina.

1605. In the mineral waters of Porta, Berzelius has discovered two new compounds, the CHRENIC and APOCHRENIC ACIDS; they are considered as products of the decomposition of vegetable matter in the neighbouring marshes.

1606. It was formerly believed that the various ingredients in saline mineral waters, are associated together in the form in which they are deposited by evaporation. The late Dr Murray, however, in his able treatise on mineral waters, has shewn that, in the water, they are more probably united in the state in which they produce the most soluble compounds. Thus sulphate of lime and chloride of sodium may be procured successively from water containing sulphuric and muriatic acids united with soda and lime. But, according to Dr Murray, these exist in the water before evaporation as sulphate of soda and muriate of lime, more soluble compounds of these ingredients, and more active in relation to the animal economy.

ACIDIMETRY AND ALKALIMETRY.

1607. Acidimetry and Alkalimetry are the terms applied to the processes by which the amount of uncombined or free acid or alkali in various substances is determined. Occasionally this may be done by ascertaining the specific gravity; but it is more generally effected, especially in dilute situations, by examining the neutralizing power of the acid or alkaline matter under examination. If not in solution, a given weight must be dissolved, or treated with water till all soluble matter is removed, the solution being tested in the following manner:

1608. If ACID, add crystallized aqueous bicarbonate of potassa in fine powder till it is neutral and no longer affects the vegetable blue: every 101 grains used contain 48 of potassa, which indicate when neutralized 40 of sulphuric, 37 of muriatic, 54 of nitric, or 51 of acetic acid, the greater number of acids neutralizing potassa in equivalent quantities.

1609. If alkaline, take any of the above acids previously diluted with water, and after having ascertained how much real acid it contains, as described in the preceding paragraph, add small successive quantities to the alkaline liquid till it is neutralized: then, if sulphuric acid shall have been used, which is generally preferred, every 40 grains of real acid used indicate 48 of potassa, 32 of soda, or 17 of ammonia in the alkaline solution.

DONIUM AND AMIDE.

1610. DONIUM is the name of a new metal, discovered by Mr Richardson. Dr Boase also noticed at the same time its oxide in another mineral.

1611. AMIDE.—This term has been applied to a compound supposed to contain 2 eqs. of hydrogen and 1 of nitrogen. Its composition has not been satisfactorily demonstrated, and it has not been procured in an insulated form. The term has also been used in different significations.

1612. Dr Gregory has discovered lately a new liquid, very similar to mercaptan, but much more volatile, boiling at 70°. He prepared it by the process for procuring mercaptan, substituting, however, pyroxilic spirit for alcohol.

1613. Dr Gregory has also procured, by the reaction of chloride of sulphur and ammonia, a compound which he considers composed of sulphur and nitrogen.

CHEMICAL NOMENCLATURE.

1614. The multiplicity of compounds which have of late been discovered, calls loudly for a revision of the present system of chemical nomenclature. The following view of the nomenclature of several compounds, will shew the system of nomenclature proposed by Dr Hope, founded upon the suggestion of Dr Thomson, that the Greek adverbial numerals should be employed in designating the number of eqs. of base, and the Latin in expressing the number of eqs. of acid.

Dr Hope has also proposed that the interweaving of the Greek and Latin numerals in the same word should be avoided, and that the prefixtures of proto, per, sub, and super, should be discontinued, as equally unnecessary, and apt to lead to a confusion of terms. Were these proposals generally adopted, they would certainly simplify much the present state of chemical nomenclature.

TABLE I.

Atom of Metal.		Atom of Oxygen.	Oxide.
1	+	1	Oxide of Metal.
1	+	2	Bis Oxide of Metal.
1	+	3	Ter Oxide of Metal.
1	+	4	Quater Oxide of Metal.
2	+	1	Dis Oxide of Metal.
3	+	1	Tris Oxide of Metal.
4	+	1	Tetrakis Oxide of Metal.
2	+	3	Dis Ter Oxide of Metal.
2	+	5	Dis Quinquoxide of Metal.

TABLE II.

Oxide.		Acid Sulphuric.	Compound Salt.
1	+	1	Sulphate of Metal.
1	+	2	Bis Sulphate of Metal.
1	+	3	Ter Sulphate of Metal.
1	+	4	Quater Sulphate of Metal.
2	+	1	Dis Sulphate of Metal.
3	+	1	Tris Sulphate of Metal.
4	+	1	Tetrakis Sulphate of Metal.
Bis Oxide.			
1	+	1	Sulphate of Bis Metal.
1	+	2	Bis Sulphate of Bis Metal.
2	+	1	Dis Sulphate of Bis Metal.
Ter Oxide.			
1	+	1	Sulphate of Ter Metal.
1	+	2	Bis Sulphate of Ter Metal.
1	+	3	Ter Sulphate of Ter Metal.
2	+	1	Dis Sulphate of Ter Metal.
Dis Oxide.			
1	+	1	Sulphate of Dis Metal.
1	+	2	Bis Sulphate of Dis Metal.
2	+	1	Dis Sulphate of Dis Metal.
Tris Oxide.			
1	+	1	Sulphate of Tris Metal.
1	+	2	Bis Sulphate of Tris Metal.
2	+	1	Dis Sulphate of Tris Metal.
Dis Ter Oxide.			
1	+	1	Sulphate of Dis Ter Metal.
1	+	2	Bis Sulphate of Dis Ter Metal.
2	+	1	Dis Sulphate of Dis Ter Metal.

FRIGORIFIC MIXTURES WITHOUT ICE OR SNOW.

(The figures before the substances used refer to parts by weight.)

1. 4 Diluted sulphuric acid* and 5 Cr. sulphate of soda,	From	50 to 3
2. 5 Common muriatic acid and 8 Cr. sulphate of soda,	...	50 to 0
3. 4 Diluted nitric acid † and 9 phosphate of soda. .	...	50 to —12
4. 1 Water and 1 nitrate of ammonia, ...	...	50 to 4
5. 16 Water, 5 nitrate of potassa, 5 muriate of ammonia, .	...	50 to 10
6. 4 Diluted nitric acid, 2 nitrate of potassa, 4 muriate of ammonia, and 6 sulphate of soda, ...	...	50 to —10

FRIGORIFIC MIXTURES WITH ICE OR SNOW.

1. 3 Snow and 2 diluted sulphuric acid, . . .	From	32 to —23
2. 7 Snow and 4 diluted nitric acid, ...	...	32 to —30
3. 8 Snow and 5 muriatic acid, ...	...	32 to —27
4. 2 Snow and 1 chloride of sodium, ...	...	60 to — 9
5. 5 Snow, 2 chloride of sodium, and 1 muriate of ammonia,	...	60 to —12
6. 2 Snow and 3 crystallized muriate of lime, . .	...	32 to —50
7. 1 Snow and 3 crystallized muriate of lime, . .	...	—40 to —73
8. 8 Snow and 10 diluted sulphuric acid, . . .	...	—68 to —91

* Composed of equal weights of strong sulphuric acid and water; allowed to cool before using it.
† Prepared by mixing one part of water and two of strong nitric acid.

INDEX.

FINIS.

PRINTED BY NEILL AND CO. OLD FISHMARKET.

Made in the USA
Monee, IL
04 November 2026

67385005R00125